High-k materials in Dynamic Random Access Memories (DRAM)

Piotr Dudek

High-k materials in Dynamic Random Access Memories (DRAM)

Atomic scale engineering of HfO2-based dielectrics for future DRAM applications

Südwestdeutscher Verlag für Hochschulschriften

Impressum/Imprint (nur für Deutschland/only for Germany)
Bibliografische Information der Deutschen Nationalbibliothek: Die Deutsche Nationalbibliothek verzeichnet diese Publikation in der Deutschen Nationalbibliografie; detaillierte bibliografische Daten sind im Internet über http://dnb.d-nb.de abrufbar.
Alle in diesem Buch genannten Marken und Produktnamen unterliegen warenzeichen-, marken- oder patentrechtlichem Schutz bzw. sind Warenzeichen oder eingetragene Warenzeichen der jeweiligen Inhaber. Die Wiedergabe von Marken, Produktnamen, Gebrauchsnamen, Handelsnamen, Warenbezeichnungen u.s.w. in diesem Werk berechtigt auch ohne besondere Kennzeichnung nicht zu der Annahme, dass solche Namen im Sinne der Warenzeichen- und Markenschutzgesetzgebung als frei zu betrachten wären und daher von jedermann benutzt werden dürften.

Coverbild: www.ingimage.com

Verlag: Südwestdeutscher Verlag für Hochschulschriften GmbH & Co. KG
Heinrich-Böcking-Str. 6-8, 66121 Saarbrücken, Deutschland
Telefon +49 681 37 20 271-1, Telefax +49 681 37 20 271-0
Email: info@svh-verlag.de

Approved by: Cottbus, BTU, Diss. , 2011

Herstellung in Deutschland:
Schaltungsdienst Lange o.H.G., Berlin
Books on Demand GmbH, Norderstedt
Reha GmbH, Saarbrücken
Amazon Distribution GmbH, Leipzig
ISBN: 978-3-8381-3018-7

Imprint (only for USA, GB)
Bibliographic information published by the Deutsche Nationalbibliothek: The Deutsche Nationalbibliothek lists this publication in the Deutsche Nationalbibliografie; detailed bibliographic data are available in the Internet at http://dnb.d-nb.de.
Any brand names and product names mentioned in this book are subject to trademark, brand or patent protection and are trademarks or registered trademarks of their respective holders. The use of brand names, product names, common names, trade names, product descriptions etc. even without a particular marking in this works is in no way to be construed to mean that such names may be regarded as unrestricted in respect of trademark and brand protection legislation and could thus be used by anyone.

Cover image: www.ingimage.com

Publisher: Südwestdeutscher Verlag für Hochschulschriften GmbH & Co. KG
Heinrich-Böcking-Str. 6-8, 66121 Saarbrücken, Germany
Phone +49 681 37 20 271-1, Fax +49 681 37 20 271-0
Email: info@svh-verlag.de

Printed in the U.S.A.
Printed in the U.K. by (see last page)
ISBN: 978-3-8381-3018-7

Abstract

Modern dielectrics in combination with appropriate metal electrodes have a great potential to solve many difficulties associated with continuing miniaturization process in the microelectronic industry.

One significant branch of microelectronics incorporates dynamic random access memory (DRAM) market. The DRAM devices scaled for over 35 years starting from 4 kb density to several Gb nowadays. The scaling process led to the dielectric material thickness reduction, resulting in higher leakage current density, and as a consequence higher power consumption. As a possible solution for this problem, alternative dielectric materials with improved electrical and material science parameters were intensively studied by many research groups. The higher dielectric constant allows the use of physically thicker layers with high capacitance but strongly reduced leakage current density.

This work focused on deposition and characterization of thin insulating layers. The material engineering process was based on Si cleanroom compatible HfO_2 thin films deposited on TiN metal electrodes. A combined materials science and dielectric characterization study showed that Ba-added HfO_2 ($BaHfO_3$) films and Ti-added $BaHfO_3$ ($BaHf_{0.5}Ti_{0.5}O_3$) layers are promising candidates for future generation of state-of-the-art DRAMs. In especial a strong increase of the dielectric permittivity k was achieved for thin films of cubic $BaHfO_3$ (k~38) and $BaHf_{0.5}Ti_{0.5}O_3$ (k~90) with respect to monoclinic HfO_2 (k~19). Meanwhile the CET values scaled down to 1 nm for $BaHfO_3$ and ~0.8 nm for $BaHf_{0.5}Ti_{0.5}O_3$ with respect to HfO_2 (CET=1.5 nm). The Hf^{4+} ions substitution in $BaHfO_3$ by Ti^{4+} ions led to a significant decrease of thermal budget from 900°C for $BaHfO_3$ to 700°C for $BaHf_{0.5}Ti_{0.5}O_3$.

Future studies need to focus on the use of appropriate metal electrodes (high work function) and on film deposition process (homogeneity) for better current leakage control.

Acknowledgments

I would like to acknowledge the people whose support and team work made this thesis possible.

It is a great pleasure to thank my supervisors, Prof. Hans-Joachim Müssig, Prof. Dieter Schmeisser and Prof. Ehrenfried Zschech for helping me with the preparation of this thesis and fruitful discussions during the three years of dissertation.

I would like to express my deepest gratitude to my wonderful colleagues from the Materials Research and Technology Departments at IHP, especially to Dr. Grzegorz Lupina, Dr. Thomas Schröder, Hans–Jürgen Thieme, Dr. Jarek Dabrowski, M.Sc. Grzegorz Kozlowski, Dr. Gunther Lippert, Dipl-Ing. Ronny Schmidt, Dr. Peter Zaumseil, Dr. Olaf Seifarth, Dr. hab. Ch. Wenger and Dr. Ioan Costina for sharing their knowledge and free time with me and for fruitful cooperation during my stay at IHP.

I express my special thanks to my colleagues and supervisors, Dr. Grzegorz Lupina and Dr. Thomas Schroeder, for their help with writing the thesis and numerous revisions made to this work and for our know-how conversations and meetings.

I would like to thank my BTU colleague who helped me within the data evaluation and participated in numerous discussions about the AFM topics: Dr. Eng. Krzysztof Kolanek.

I would like to thank my wonderful parents for driving me through the entire education process during those years.

Die Doktorarbeit wurde innerhalb des durch das Bundesministerium für Bildung und Forschung finanzierten MEGAEPOS (Metall-Gate-Elektroden und epitaktische Oxide als Gate-Stacks für zukünftige CMOS-Logik- und Speichergenerationen) Verbundprojektes abgeschlossen. Das IHP-Teilvorhaben war die Herstellung und Analyse alternativer dielektrischer Schichten für künftige CMOS-Bauelemente (FKZ: 13N9261).

Publications

Part of this work was published by the author in the following articles.

1. Atomic–scale engineering of future high–k DRAM dielectrics: the example of partial Hf substitution by Ti in BaHfO$_3$
P. **Dudek**, G. Lupina, G. Kozlowski,J. Bauer, O. Fursenko, J. Dabrowski, R. Schmidt, G. Lippert, H-J. Müssig, D. Schmeißer, E. Zschech and T. Schroeder, *J. Vac. Sci. Technol. B 29, 01AC03-01AC03-7 (2011)*

2. Basic Investigation of HfO$_2$ based Metal-Insulator-Metal Diodes
P. **Dudek**, R. Schmidt, M. Lukosius, G. Lupina, C. Wegner, A. Abrutis, M. Albert, K. Xu, A. Devi, *submitted to Thin Solid Films 519, 5796-5799 (2011)*

3. Characterization of group II hafnates and zirconates for metal-insulator-metal capacitors
G. Lupina, O. Seifarth, P. **Dudek**, G. Kozlowski, J. Dabrowski, H-J. Thieme, G. Lippert, T. Schroeder, H-J. Müssig, *accepted for publication in Phys.Stat.Sol. B (2010)*

4. Perovskite BaHfO$_3$ dielectric layers for dynamic random access memory storage capacitor applications
G. Lupina, J. Dabrowski, P. **Dudek**, G. Kozlowski, M. Lukosius, Ch. Wenger, H-J. Müssig, *Adv. Eng. Mat., 11, 4 (2009)*

5. Deposition of BaHfO$_3$ dielectric layers for microelectronic applications by pulsed liquid injection MOCVD
G. Lupina, M. Lukosius, C. Wenger, P. **Dudek**, G. Kozlowski, H.-J. Müssig, A. Abrutis, R. Galvelis, T. Katkus,Z. Saltyte, V. Kubilius, *Chem. Vap. Dep., 15, 167 (2009)*

6. Hf-and Zr-based alkaline earth perovskite dielectrics for memory applications
G. Lupina, O. Seifarth, G. Kozlowski, P. **Dudek**, J. Dabrowski, G. Lippert, H.-J. Müssig, *Microelectronic Engineering, 86, 1842 (2009)*

7. Group II hafnate and zirconate high-k dielectrics for MIM storage capacitors in DRAM - the defect issue

J. Dąbrowski, **P. Dudek**, G.Kozlowski, G. Lupina, G. Lippert, R. Schmidt, Ch. Walczyk, and Ch. Wegner, *ESC Trans., 25, 219 (2009)*

8. Dielectric properties of Hf and Zr based alkaline earth perovskite layers

G. Lupina, **P. Dudek**, G. Kozlowski, J. Dąbrowski, G. Lippert, H-J. Müssig, and T. Schroeder, *ESC Trans., 25, 147 (2009)*

9. Thin $BaHfO_3$ high-k dielectric layers on TiN for memory capacitor applications

G. Lupina, G. Kozłowski, J. Dabrowski, Ch. Wenger, **P. Dudek**, P. Zaumseil, G. Lippert, Ch. Walczyk, and H.-J. Müssig, *Appl. Phys. Lett., 92, 062906 (2008)*

10. Dielectrics Characteristics of Amorphous and Crystalline $BaHfO_3$ High-k Layers on TiN for Memory Capacitor Applications

G. Lupina, G. Kozlowski, **P. Dudek**, J. Dabrowski, Ch. Wenger, P. Zaumseil, G. Lippert, H.-J. Müssig, *9th Conference on Ultimate Integration on Silicon, ULSI 2008, Udine, March 12-14, 2008, Italy*

4

List of Terms

AFM	atomic force microscopy
ALD	atomic layer deposition
ASF	atomic sensitivity factor
AVD	atomic vapour deposition
BG	band gap
BL	bit line
CET	capacitance equivalent thickness
CMOS	complementary metal-oxide-semiconductor
C-V	capacitance-voltage
C-AFM	conductive AFM
CBE	conduction band edge
CBM	conduction band minimum
CBO	conduction band offset
COB	capacitor-over-bit-line
CUB	capacitor-under-bit-line
CVD	chemical vapour deposition
C_{BL}	capacitance on the bit line
C_{TOT}	total capacitance
DRAM	dynamic random access memory
DT	deep trench
E	electric field
E_A	activation energy
E_B	binding energy
E_F	Fermi energy
E_g	band gap energy
E_{KIN}	kinetic energy
ε_r	dielectric permittivity
FeRAM	ferroelectric random access memory
GIXRD	grazing incidence x-ray diffraction
IC	integrated circuit

IMFP	inelastic mean free path
ITRS	international technology roadmap for semiconductors
J-V	current-voltage
k	dielectric constant
MBD	molecular beam deposition
MIM	metal-insulator-metal
MRAM	magnetic RAM
NVM	non-volatile memory
PCRAM	phase change RAM
P	polarization
PSD	position sensitive diode
PVD	physical vapour deposition
RBS	Rutherford backscattering spectroscopy
RMS	root mean square
RTA	rapid thermal annealing
SE	spectroscopic ellipsometry
SOS	spin-orbit-splitting
SR-XAS	synchrotron-radiation XAS
STM	scanning tunnelling microscopy
TEM	transmission electron microscopy
UPS	ultraviolet photoelectron spectroscopy
VBM	valence band maximum
VBO	valence band offset
WL	word line
WF	work function
XAS	X-ray absorption spectroscopy
XPS	X-ray photoelectron spectroscopy
XRR	X-ray reflectometry
XRD	X-ray diffraction

Contents

Chapter 1

Overview

1.1 Goal of the study

Since the dawn of the electronic era, memory or storage devices have been an integral part of electronic components. As the electronic industry matured and moved away from the vacuum tubes to semiconductor devices, research in the field of semiconductor memories intensified as well. The semiconductor memory industry evolved and prospered along with computers revolution [1]. In 1970, the newly formed Intel company released the "1103", the first dynamic random access memory (DRAM) chip and by 1972 it was the best selling semiconductor memory on the market defeating magnetic core type memory (Fig. 1.1).

Figure 1.1: Intel "1103" first DRAM chip [2].

High density and low cost of DRAMs have earned them a predominant role in computer main memories. During the last four decades, the number of DRAM chips has increased four times every three years and the cost per bit has declined by the same factor [3].

The DRAM scaling progressed successfully for over 35 years from the 4 kb density up to several Gb nowadays, resulting in an increase of chip complexity by many orders of magnitude, i.e. memory technology within IBM products has undergone a 280000-fold increase in density, 20000 times decrease in power per bit, and a 10 to 100 times increase in speed during the last twenty-five years, resulting in a 650-fold reduction of the cost per bit of memory (Fig. 1.2) [4, 5].

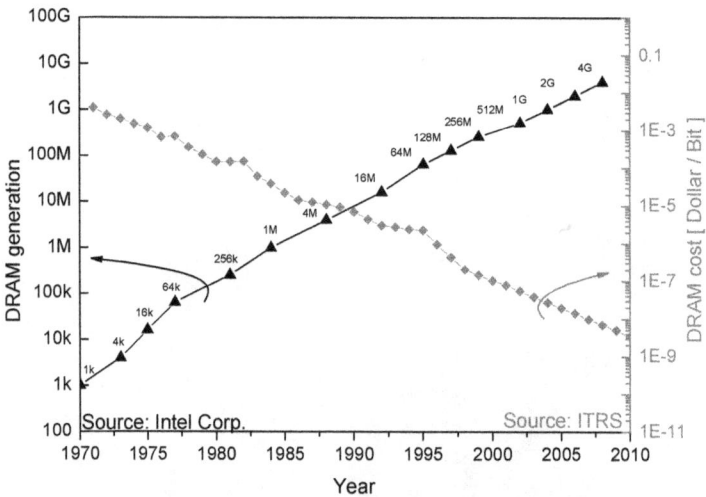

Figure 1.2: Time evolution of DRAM generation and DRAM cost.

Memories are used for writing, storing and retrieving large amounts of data, which represent either information or software instructions that are coded in combination of binary digits "0" and "1" and can be further used by the processor unit for specific manipulation.

To the main attributes of DRAM memory belong cost, reliability, density, speed and power consumption. Progressively the electronic devices become smaller, faster and cheaper, as the rapid scaling process proceeds.

However, reducing device dimensions leads also to the thickness reduction of electrical barriers in the device. These barriers with reduced thickness can not provide sufficient insulation anymore, resulting in higher leakage current density and higher power consumption. In DRAM, one of the main contributions to the higher power dissipation is the leakage through the storage capacitor dielectric. As a remedy for this problem, alternative dielectric materials are studied. These materials are called high-k dielectrics. The term high-k refers to a material with a high dielectric constant k (also called dielectric permittivity ε_r) as compared to silicon dioxide (SiO_2). High-k dielectrics are required as means to provide a substantially thicker (physical thickness) dielectric for reduced leakage and improved capacitance, in other words for enabling continued equivalent oxide thickness scaling, and hence high performance of microelectronic memory.

The goal of this thesis is the preparation and characterization of new dielectric materials with respect to their application in MIM storage capacitors of future DRAM generations.

1.2   Organisation of the thesis

The thesis is organized as follows:

Chapter 2 takes up with reader's introduction to the topic. After recalling the main microelectronic memory types, it reviews the DRAM operation principle. In the next step the miniaturization approach is presented with focus on the replacement of the current insulator in DRAM storage capacitors by an alternative one. The main requirements for potential candidates are listed in detail.

Chapter 3 provides a general overview of film deposition and characterization methods used in this study. Description of deposition equipment and measurement techniques is provided.

Chapter 4 summarizes the results of physical and electrical characterization of high-k materials developed in this study. It is divided into two main parts. The first part is focused on Hf-based alkaline earth perovskites formed by addition of BaO to HfO_2. The second part is focused on the further engineering of ternary compounds, by addition of TiO_2. The main focus is put on $BaHf_{1-x}Ti_xO_3$.

Chapter 5 summarizes obtained results and states suggestions for future work.

Chapter 2

Introduction

2.1 Memory types

In the past, data storage used several memory technologies, i.e. magnetic tapes, hard disks, floppy disks, core memories, optical discs and semiconductor memories. Magnetic and optical media belong to the nonvolatile memory (NVM) family as well as other, non-charge storage types including Ferroelectric Random Access Memory (FeRAM), Magnetic RAM (MRAM) and Phase Change RAM (PCRAM) [6]. Within the NVM group, once the information is written, it is retained permanently, even if the power supply is disconnected. In opposite, volatile memories loose the information after the power is switched off. Some loose information if it is not refreshed over certain periods of time [3].

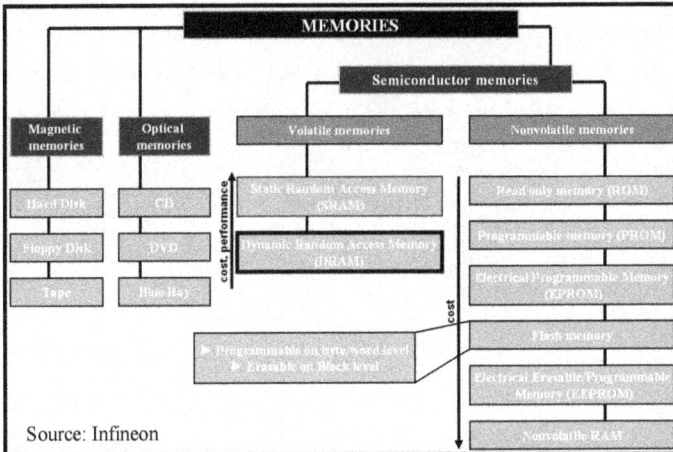

Fig. 2.1 Memory types [after T. Mikolajick].

13

An exact illustration of memory types is given in Fig. 2.1. To the nonvolatile group belongs read only memory, usually known by its acronym ROM. Because data stored in ROM can not be modified, it is mainly used to distribute firmware. Programmable ROM (PROM) is a form of digital memory where the setting of each bit is locked by a fuse or antifuse. Such units are used to permanently store programs. The difference with respect to a ROM is that the programming is applied after the devices are constructed and they are mostly used in electronic dictionaries. The memory can be programmed only once and it is an irreversible process. Erasable programmable ROM (EPROM) is a NVM type that consists of floating-gate transistors individually programmed by an electronic device which supplies higher voltage than normally used in digital circuits. The EPROM can be erased by exposure to strong UV light (253.7 nm). In contrast, electrically erasable programmable ROM (EEPROM, often written as E^2PROM) is a type of memory used in electronic devices to store small amounts of data in the case of power removal, i.e. device configuration. When larger amounts of data have to be stored, a specific type of EEPROM is used, namely flash memory. It can be electrically erased and reprogrammed. The last one from this group, NVRAM gathers all types of RAMs which do not lose their data upon power removal. This is in contrast with DRAM which belongs to the volatile memory group [8]. SRAM is a type of memory where the word "static" indicates, that, unlike DRAM, it does not need to be periodically refreshed. SRAM uses bistable latching circuitry to store each bit, however it is still volatile in the conventional sense that data is lost when the memory is not powered. The focus of this work is on the last semiconductor memory type, namely DRAM.

14

2.2 DRAM

In the following subchapters, the DRAM working principle is described. It is followed by capacitor structure and capacitor physics.

2.2.1 DRAM working principle

At present, a so called one-transistor-one-capacitor (1T-1C) structure is usually used as a memory component of the DRAM unit cell [9]. Fig. 2.2 shows in the left upper corner an access transistor (T) and a capacitor (C) cell (enlarged on the right side) with top and bottom plate and ZrO_2 dielectric between them. The storage element is connected via contact 2 and contact 1 (bit line - BL) to the transistor (Fig.2.2 left, down) with $CoSi_2$ metallization for electrode contacts (word line - WL).

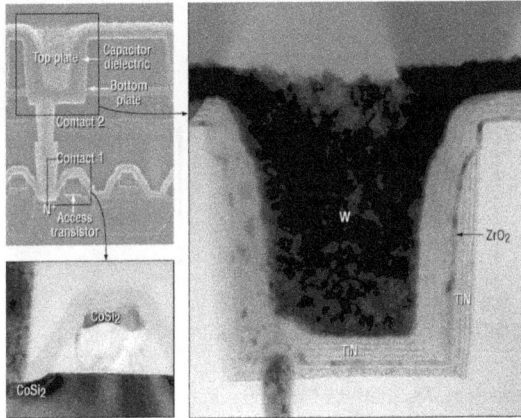

Figure 2.2: 1T/1C DRAM cell unit [10].

In the following, the processes of writing to and reading from a DRAM cell are shortly described.

Writing

The word line applies a write voltage such that the access transistor is in the "on" state and behaves like a metal wire. During the writing process, the BL applies a voltage which is dependent on the information to be stored (Fig.2.3). It results in charging the storage capacitor.

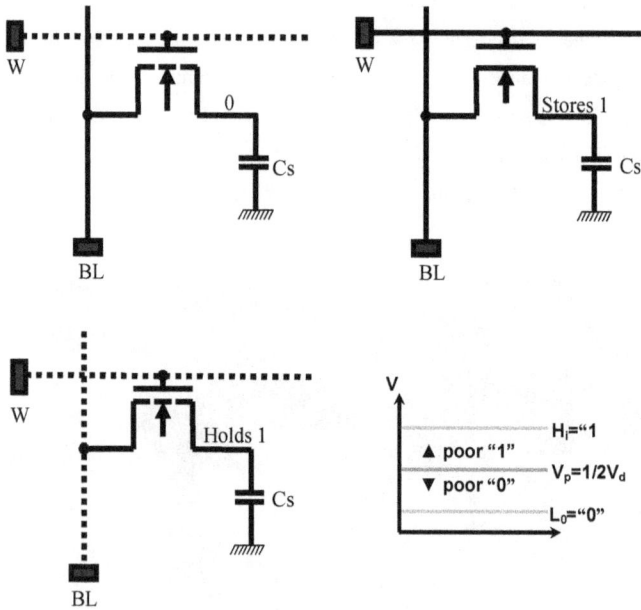

Figure 2.3: Schematic representation of writing process; "0" and "1" state during writing process (bottom, right).

If the applied voltage is slightly higher than the half of V_{DD} (V_{DD} stands for positive power supply and "D" denotes that the supply is connected to the drains), a logical "1" is stored in the cell. If the value is slightly lower than $1/2V_{DD}$, a logical "0" is saved. The main advantage of the half-voltage method is that the established insulator field is twice smaller than in V_{DD} method what contributes to a better reliability [48].

Reading

In the reading process, the BL is precharged to the voltage $V_{BL}=V_{DD}/2$ whereas the word line (WL) remains closed (Fig. 2.4). In the next step WL opens and voltage is applied to transistor gate. The channel is conducting and electric charge flows through the source and drain. If the storage capacitor C_s held a logical state "1" (called also "high" state), then the potential on the BL will be equal $V_{BL}+\Delta V_S$ (V_S stands for voltage on storage element S). The potential on the BL capacitor (C_{BL}) will be slightly higher than the applied one, and the charge will be transferred in the direction C_S-C_{BL}. The opposite situation suggests a logical state "0" at the storage capacitor because C_{BL} will be depleted in charge. It has to be mentioned that reading procedure is a destructive process; released charge is lost [7].

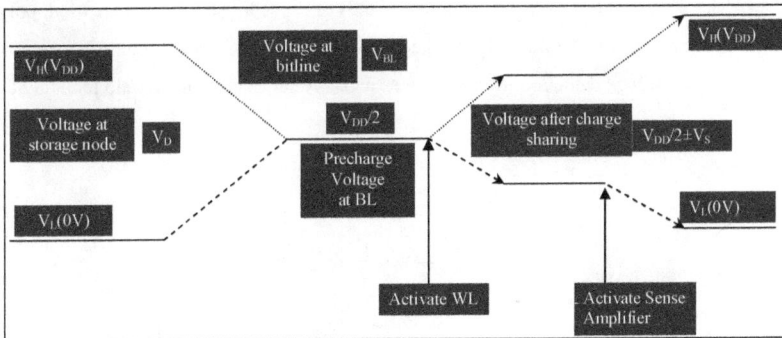

Figure 2.4: Schematic representation of DRAM reading process.

2.3 DRAM capacitor structure

The geometric structure of the storage capacitor has been evolving throughout the years of scaling [11, 31]. Different forms have been applied to obtain improvements in capacitance density. One can distinguish two major designs: planar and three dimensional (3D) structures [3]. The working principle is the same in each.

Fig. 2.5 shows how the DRAM capacitor structure evolved through the shrinking technology nodes. Colors represent different functional materials; red stands for metal gate (WL), blue for metal plate of the capacitor, green for the dielectric and grey for p-Si. BL is shown on the top, along the unit cell.

Planar capacitor design:

As an example 16-KBit to 1 MBit DRAMs in early 70s to the mid- 80s belonged to the planar device cell family. This structure based on a transfer device (n- or p-channel MOSFET) and a capacitor placed horizontally along the transfer side and occupying typically 30% of the DRAM cell area. Two kinds of capacitor structures were used; poly-insulator-silicon (PIS) and poly-insulator-poly (PIP). The change from PIS to PIP reduced the silicon depletion layer, acting as parasitic series capacitance and reducing the total capacitance of the unit cell.

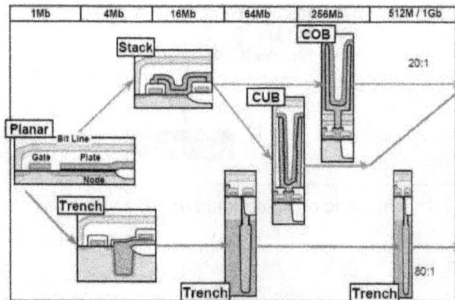

Figure 2.5: DRAM capacitor structure, schematic depiction; colors in the figure: red (print dark grey) stands for metal gate (WL), blue (print black) for metal plate of the capacitor, green (print light grey) for the dielectric and grey for p-Si [48-49].

18

3D capacitor design:

Fabrication of 4-Mbit DRAM required the introduction of 3D structures to follow the scaling trend without effective capacitor area reduction. One of the ideas was to position the capacitor within a deep trench (DT) (Fig. 2.5) [12, 48]. Trench developed very quickly by increasing the depth/width (D/W) ratio and using thereby more efficient the silicon substrate area.

TRENCH

Within the 16 Mb up to the 256 Mb DRAM cells, density enhancing innovations focused on techniques as shallow trench isolation (STI) [13], BL contact borderless to word line [14], and self aligned buried strap [15]. 256 Mb DRAM generation manufactured in 2001 had a cell size of approximately 0.16 μm^2 and a minimum pitch o 0.28 μm^2 [16, 17]. Furthermore, 70 nm and a 58 nm technology were reported [18, 19] as well as 40 nm DRAM cell [20]. A trench aspect ratio of 90 was presented for the 70 nm technology node and an aspect ratio of 120 within the 40 nm node. By the year 2007, scaling of DT structures became problematic. Major issues included technological difficulties with further increasing of the aspect ratio and uniform filling with electrode and dielectric materials, as well as very high thermal budget which these materials had to withstand (~1000 °C); for these reasons the stack architecture became most popular.

STACK

The second type of 3D structure is the stack capacitor. As it was introduced, the requirement for the dielectric constant of the material changed towards higher values in order to keep the same capacitance thickness. For sub-100 nm technologies, a cylinder capacitor structure utilizing the inner and outer cylinder surface area and a cup capacitor structure was proposed. It delivers only half of the capacitor area at the same aspect ratio as DT but is mechanically more robust [21, 22]. Two geometric structures were distinguished here – capacitor-over-bit-line (COB) and capacitor-under-bit-line (CUB). Both solutions present the same major concern: keep the same aspect ratio for contacts/vias whatever the generation for manufacturability is [51]. Stack capacitor is the mainstream technology nowadays.

2.4   DRAM capacitor physics

To understand the physics of DRAM, one has to focus on the dielectric between the metallic conductors, a polarisable insulating material.

M. Faraday found that placing an insulating material between metal plates connected to a voltage source, increased the charge on those plates by a factor of ε_r, called dielectric constant or relative permittivity of the material [50]. Dielectric medium has no free charges, in other words, positive and negative charges average over the scale of an atom, however an external electric field $\vec{E}$ causes them to displace, forming an induced dipole. The electric field between electrodes stores energy until they are connected. If it occurs, the field collapses and charges neutralize.

The ability of a passive element to store these charge carriers is called capacitance and describes the number of Coulombs of charge that can be placed on a capacitor plate per unit of voltage applied [23]:

$$Q = C \cdot V \tag{2.1}$$

Using Gauss's law we can determine the electric field between the plates. The electric flux through a closed surface A, surrounding the region V is proportional to the electric charge Q_A in this region:

$$\Phi = \oint_A \vec{E} \cdot d\vec{A} = \frac{1}{\varepsilon_0} \int_V \rho \, dV = \frac{Q_A}{\varepsilon_0} \tag{2.2}$$

where vector $d\vec{A}$ is a surface vector, $\vec{E}$ electric field, ρ is the total electric charge density and ε_0 is the dielectric permittivity of vacuum. The electric field is then equal to the electrical flux divided by the area:

$$E = \frac{\Phi}{A} = \frac{Q_A}{A \cdot \varepsilon_0} = \frac{C \cdot V}{A \cdot \varepsilon_0} \tag{2.3}$$

In homogenous parallel plate capacitor electric field, E=V/d, where d stands for thickness (or distance between plates), therefore:

$$E = \frac{V}{d} = \frac{C \cdot V}{A \cdot \varepsilon_0}$$ (2.4)

and finally
$$C = \frac{A \cdot \varepsilon_0}{d}$$ (2.5)

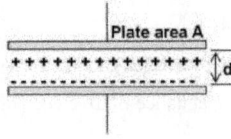

Figure 2.6: Parallel plate capacitor

To complement the case with dielectric material between metal plates the equation takes the form:

$$C = \frac{A \cdot \varepsilon_0 \cdot \varepsilon_r}{d}$$ (2.6)

where ε_r stands for the dielectric permittivity of the material and is often replaced by k. In DRAM, if the capacitor area A is shrinking, the term C has to remain constant, nominally 30-40fF/storage cell [11]. This is because the charge in the capacitor must be large enough to generate a BL voltage change that can be reliably sensed, including compensating for various noise sources (radiation, leakage current and electrical imbalances between pairs of BLs) [24, 25]. To keep C constant with decreasing A, a reduction in d is required.

Decreasing the dielectric thickness can result in an increase of the leakage current density due to direct tunneling (DT). According to the quasi-classical Wenzel-Kramers-Brillouin (WKB) approximation, DT current can be expressed as:

$$J = J_0 \cdot \left\{ \overline{\varphi} \cdot \exp \cdot (-A \cdot \sqrt{\overline{\varphi}}) - \left(\overline{\varphi} + eV \right) \exp \cdot (-A \cdot \sqrt{\overline{\varphi} + eV}) \right\}$$ (2.7)

$$J_0 = \frac{e}{2 \cdot \pi \cdot h \cdot d^2} \quad \text{and} \quad A = 4 \cdot \pi \cdot d \cdot \sqrt{\frac{2 \cdot m^* \cdot e}{h}}$$

21

where $\bar{\varphi}$ stands for mean dielectric barrier height above Fermi level, d for oxide thickness, m* for effective electron mass, e for electron charge, h for Planck constant and V for voltage across the film [26]. As a consequence, it follows from the DT formula that the DT current increases exponentially with decreasing dielectric film thickness d. Therefore, dielectric thickness could not be reduced infinitely to satisfy leakage current and high capacitance density requirements at the same time. Therefore, to enable further miniaturization, another solution must be targeted: alternative dielectrics with higher dielectric permittivity ε_r [34, 44].

To compare electrical performance of higher ε_r dielectrics with that of SiO_2, a term named CET was introduced. Capacitance Equivalent Thickness (CET) is determined by:

$$CET = \frac{A \cdot \varepsilon_{SiO_2} \varepsilon_0}{C_{meas}} \qquad (2.8)$$

where C_{meas} is the measured capacitance at defined voltage and ε_{SiO2} the SiO_2 relative permittivity. CET indicates how thick would be SiO_2 film in order to provide the same capacitance as the high-k material and is extracted from electrical data.

During material deposition on metallic electrode, a parasitic interface layer can appear which adds a parasitic capacitance contributing to the total capacitance C_{TOT} (Fig. 2.7):

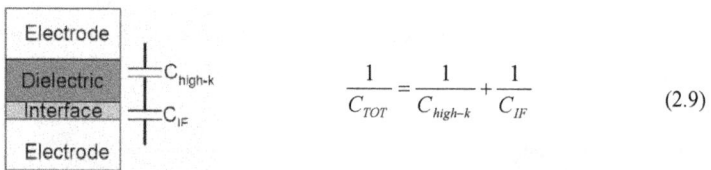

$$\frac{1}{C_{TOT}} = \frac{1}{C_{high-k}} + \frac{1}{C_{IF}} \qquad (2.9)$$

Figure 2.7: MiM structure with performance limiting interfacial layer[1].

Accordingly, the equation for CET changes its form to:

---

[1] The interface from top electrode is neglected under the assumption that Au/Pt does not react with the dielectric layer.

$$CET = A \cdot \varepsilon_{SiO_2} \cdot \varepsilon_0 \cdot \left[\frac{1}{C_{TOT}} \right] = A \cdot \varepsilon_{SiO_2} \cdot \varepsilon_0 \cdot \left[\frac{1}{C_{high-k}} + \frac{1}{C_{IF}} \right] \qquad (2.10)$$

$$CET = \frac{\varepsilon_{SiO_2}}{\varepsilon_{high-k}} \cdot t_{high-k} + \frac{\varepsilon_{SiO_2}}{\varepsilon_{IF}} \cdot t_{IF} \qquad (2.11)$$

If the denominator ε_{high-k} of the first part in equation (2.11) is high, than the ratio $\varepsilon_{SiO2}/\varepsilon_{high-k}$ might be neglected in comparison with the second part where the denominator is small [27]. The total capacitance is then determined by the interface layer. This is why great care must be taken to avoid interfacial layer formation between high-k dielectrics and metal electrodes[1].

According to the International Technology Roadmap for Semiconductors (ITRS), CET values scale with technology node, starting with 1.2 nm in 2007 and ending with less than 0.2 nm in 2022. The maximum tolerable leakage current density is defined as 10^{-8} A/cm^2 at 0.5V [6]. Regarding equation (1.7) and assuming ideal dielectric (without defects and interface) and tunneling barrier of 2.3 eV, with effective mass 0.2 m$_e$ (parameters in agreement with literature with respect to HfO$_2$ [52]), and combining those parameters with CET, one can estimate (Fig. 2.8) the predicted dielectric permittivity for the year 2012 according to the roadmap to be about 50 [28]. In case of "real" materials which are very likely to contain defects, even higher dielectric constants will be required.

---

[1] In this work, the impact of quantum effects is not included: the influence on the CET value is in the region of 0.05 nm which is within the marginal error of our measurement.

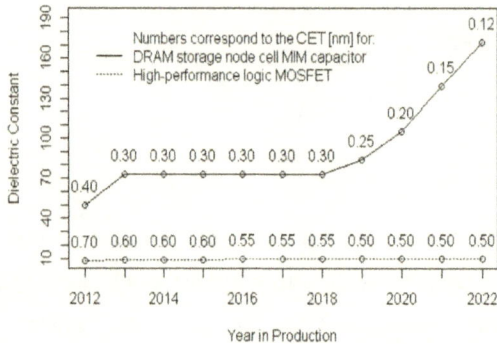

Figure 2.8: Dielectric constant vs. year, DRAM storage node cell of MIM capacitor shown by solid line, values on the plot refer to corresponding CET values [28].

2.5 DRAM dielectric

First DRAM's were based on SiO_2 and manufactured in the silicon-insulator-silicon (SIS) form. Then SiO_2/Si_3N_4 (ON/ONO, oxide-nitride and oxide-nitride-oxide, respectively) was introduced. The SiO_2 had a relative dielectric constant of 3.9 while the Si_3N_4 about 6-8 depending on stoichiometry. From 2001 onwards, radical dielectric material changes were introduced. Si_3N_4 is abandoned in stack architectures due to its high thermal budget but it remains mainstream dielectric in DT caps [29]. MIS stacked-capacitors using high-k materials and metallic top electrodes are reported with 1 Gb chips using Ta_2O_5 and 4 Gb using Al_2O_3 in 100 nm technology. DRAM then enters sub-100 nm technology CMOS nodes, focusing on design and application issues. New high-k dielectric materials, as HfO_2 and ZrO_2, with low thermal budget, are introduced to replace Ta_2O_5 in stack capacitors, along with the introduction of MIM cells [30].

Figure 2.9 presents the memory ground rule vs. CET for stack and trench technologies below 90nm. Trench DRAM utilized ON in the SIS structure up to 65 nm technology node. In order to scale it below 50 nm, an introduction of MIS construction and new material (HfSiO) was necessary. The 90 nm ground rule stack capacitor based on HfAlO in MIM configuration. As further scaling proceeds, ZrAlO will be in focus until sub-40 nm technology node. Continuing progress will demand admixture of TiO_2 to binary compounds, resulting in ternary dielectrics. There are no known solutions concerning dielectric materials for the future 28 nm technology node.

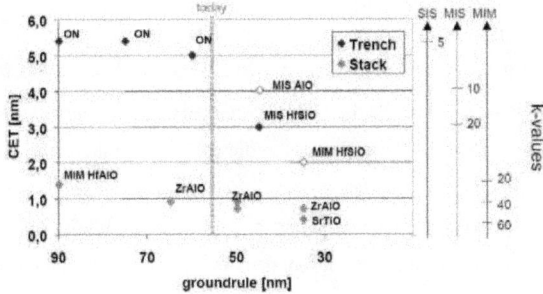

Figure 2.9: Design rule vs. capacitance equivalent thickness. DRAM structure, dielectric material as well as the electrode material change [48].

Therefore, making use of its well—established Si CMOS compatibility, we will use HfO_2 ($\varepsilon_r \sim 20$ for amorphous and monoclinic phases) as a study material and make an attempt to improve the dielectric parameters (i.e. leakage density, dielectric permittivity) in such a way that it fulfills the requirements of the future 28 nm technology.

Dielectric constant engineering

As described by the Clausius-Mossotti equation, the dielectric permittivity ε_r is defined as a function of dielectric polarizability (α) and the molar volume (V_m):

$$\varepsilon_r = \frac{1 + \frac{2}{3} \cdot 4 \cdot \pi \cdot \frac{\alpha}{V_m}}{1 - \frac{1}{3} \cdot 4 \cdot \pi \cdot \frac{\alpha}{V_m}} \qquad (2.12)$$

Hence, increasing α and decreasing V_m of the dielectric compoound, increa dielectric permittivity ε_r of the material.

25

The static dielectric permittivity is frequency dependent and in the frequency regime of interest there are two main contributions:

$$\varepsilon_{TOT} = k = k_i + k_e \qquad (2.13)$$

where k_i and k_e correspond to the lattice and electronic contribution, respectively. Atoms with a large ionic radius (i.e. high atomic number) exhibit more electron dipole response to an applied electrical field. The electronic contribution therefore, tends to increase the dielectric permittivity for higher atomic number atoms. The ionic contribution can be however larger than the electronic part in case of perovskite material, e.g. $(Ba, Sr)TiO_3$ [37]. Due to the fact that ions respond more slowly to an applied electrical field than electrons, the ionic contribution decreases at very high frequencies in the range of $\sim 10^{12}$ Hz.

Concerning the electronic polarizability plot versus ionic radius for divalent cations (Fig. 2.10), one may conclude that the electronic polarizability of Ba belongs to the highest one, showing $\alpha \approx 6.5 \ \text{Å}^3 = 6.5 \cdot 10^{-24} \text{cm}^3$ [38, 39]. Numerous literature suggests that admixture of TiO_2 can be beneficial for increasing k. TiO_2 has a large k which arises through a strong contribution from soft phonons involving Ti ions and is not exhibited by the other group IVB metal oxides [53].

Taking into account the considerations above, our approach to engineer the dielectric constant of HfO_2 will be focused on three points:

1. Adding Ba to HfO_2 to increase the electronic polarizability.

2. Obtaining crystalline $BaHfO_3$ preferably in the high symmetry cubic perovskite phase (ABO_3) to decrease V_m [40-42].

3. Substitution of a part of B-site atoms by Ti to introduce easily polarizable Ti-O bonds to further enhance dielectric constant [53].

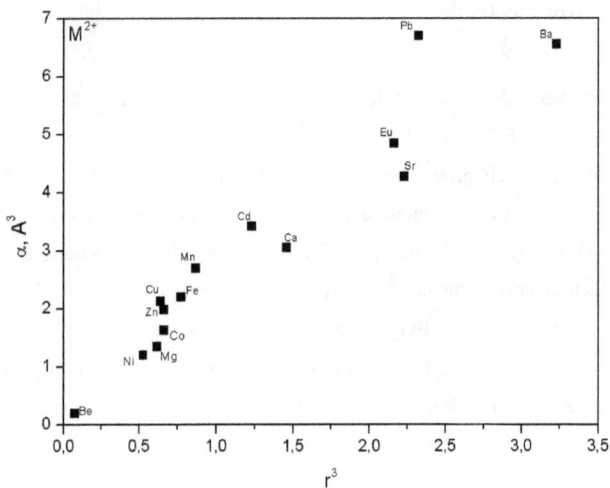

Fig. 2.10: Electronic polarizabilities ($\text{Å}^3=10^{-24}\text{cm}^3$) of divalent cations vs. (ionic radius)^3 [38].

According to the C-M equation, this should allow to obtain higher permittivity values than that of pure HfO_2.

2.6 Requirements placed on dielectrics materials for capacitors

The dielectric materials prepared according to the discussed approach, will be evaluated in this work with respect to the following criteria:

1. Dielectric constant $\varepsilon_r > 50$
2. Leakage current density $J_{leak} < 10^{-8}$ A/cm^2 at 0.5 V
3. Low thermal budget < 650 °C
4. Capacitance equivalent thickness CET < 0.7 nm

2.7 Capacitor electrode

It has been shown that dielectrics with higher dielectric constant usually show smaller band gaps (BG) [44-45]. This trend is illustrated in Fig.2.11. It can be seen that BG is roughly inversely proportional to the permittivity (more exactly $E_g \sim k^{-0,65}$ [33, 46]). The decreasing BG means a reduced energy distance between the oxides valence band maximum (VBM) defined by occupied O 2p states and the conduction band minimum (CBM) which in most of the oxides of interest is defined by metal d-states [46]. However, not only the total size of the BG is of importance. The energetic positions of CBM and VBM with respect to Fermi Energy (E_F) of the metal electrode, defining the so called conduction band offset (CBO) and valence band offset (VBO), are far more critical.

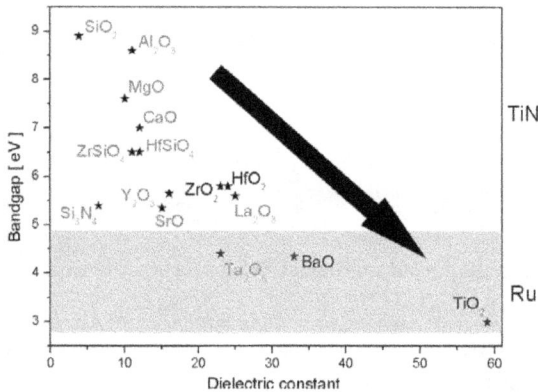

Figure 2.11: Band gap vs. dielectric constant [47].

Usually, the CBO for candidate oxides on Si or TiN is smaller than VBO [46-47]. This makes the value of CBO more critical for obtaining low leakage currents. For some oxides with very high dielectric constant, the CBO is reduced to values well below 1eV. Such barriers can not ensure ultra low leakage currents required by DRAM. A solution in this case may be the application of metal electrodes with higher work functions (WF) than that of TiN. Using this approach, the Fermi level can be shifted towards the mid-gap position of the insulator. As a result, the CBO is increased (at the expense of VBO) and leakage can be reduced. For this reason, the traditional TiN electrodes may be replaced in the future by metals with higher WF, e.g. Ru (see Fig. 2.11).

Chapter 3

Experimental methods

This chapter gives a short description of experimental methods applied in this work for the preparation and characterization of dielectric materials and metal electrodes. For a detailed description of the different materials science and electrical characterization techniques, the reader is referred to the corresponding literature.

3.1 Thin film deposition

To characterize the properties of new dielectric materials, planar metal–insulator–metal (MIM) capacitors were prepared. In the following, materials and methods used in the preparation process are described.

3.1.1 Substrates

4–inch, boron–doped (5–15 Ωcm) Si (001) wafers covered with thin TiN layers (~20 nm) were used as substrates. The TiN layers were prepared by DC magnetron sputtering of Ti in Argon/Nitrogen ambient at 300°C.

3.1.2 Dielectric material deposition

A large variety of deposition methods was applied in the research of new dielectric materials including Chemical Vapor Deposition (CVD), Atomic Layer Deposition (ALD), Atomic Vapor Deposition (AVD), Physical Vapor Deposition (PVD), Molecular Beam Deposition (MBD) and sputtering [54].

Most thin films used in microelectronic industry are deposited using CVD or ALD. Although they offer a very good control over the film stoichiometry and uniformity, expensive precursor chemistry, instrumentation costs and high reproducibility make them suitable for industrial production.

However, due to the low flexibility of CVD and ALD for research tasks, MBD by electron beam evaporation (e–beam) and effusion cell (Knudsen cell) was utilized in this study. This method offers clean, cost effective and rapid solution for primary material screening studies. In the electron evaporator, a focused electron beam heats the target (in this study: HfO_2, Ti_2O_3) and source atoms are evaporated towards the substrate. In Knudsen cell (in this study: BaO), high temperature is applied to the material by indirect ohmic heating in the effusion cell leading to material evaporation [55]. The parallel use of both sources allows stoichiometric deposition of complex compounds. The purity of materials evaporated in this study was better than 99.99%.

The dielectric deposition experiments were carried out in an ultra high vacuum (UHV) molecular beam epitaxy (MBE) system (DCA MBE 600) designed for handling 4–inch wafers. The setup consists of a load–lock, preparation and characterization chamber, shown in Fig. 3.1, connected by an UHV transfer line. In the load-lock stage, samples were degassed (200 °C for 20 min at 10^{-6} mbar) to free the surface from contaminations. During the dielectric deposition the pressure was kept around 10^{-7} mbar.

The substrate temperature was set to 400 °C during deposition and the deposition rates were typically in the range of 0.01 to 0.05 nm/s.

Fig. 3.1 DCA 600 UHV Molecular Beam Deposition chamber for preparation (left) and *in situ* characterization (XPS, UPS) (right) of thin dielectric films used in this study.

3.1.3 Post deposition treatment

Selected samples were directly after deposition annealed for 15 s using a Rapid Thermal Annealing (RTA) tool from Jipelec (Jetfirst 100) shown in Fig. 3.2. The annealing ambient was N_2 and temperatures were varied in the range from 600 °C to 900 °C.

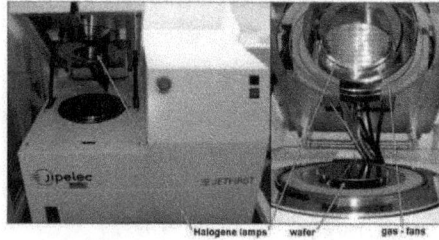

Fig. 3.2 JIPELEC JETFIRST Rapid Thermal Annealing furnace

3.1.4 Top metal electrode deposition

For top metal contact evaporation a MSBA–580 TSEW (Malz & Schmidt) metallization chamber was applied (Fig. 3.3).

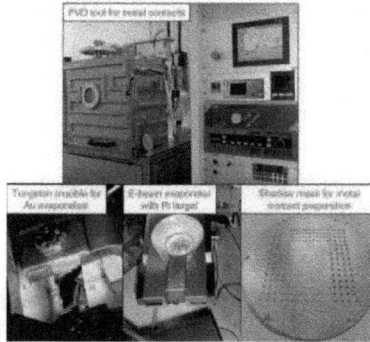

Fig. 3.3 Metallization chamber MSBA–580 TSEW (Malz & Schmidt) for 4 and 8 inch wafers for TiN, Al, Au, Pt and Ti metal contacts, shadow mask for metal contact preparation

Two types of metal electrodes were evaporated through a shadow mask: Gold (Au) and Platinum (Pt). For Au electrodes resistive thermal evaporation was applied. The material is evaporated by passing a large current through a crucible (in the boat form) which has a finite electrical resistance. For Pt contacts, an electron beam evaporation source was used [Fig. 3.3].

3.2 Characterization methods

In this subchapter, characterization methods which followed the deposition process will be described.

3.2.1 Materials science characterization

3.2.1.1 X–ray photoelectron spectroscopy

The characterization chamber shown in Fig. 3.1 allows for in situ control of the chemical composition of the dielectric layers using X–Ray Photoelectron Spectroscopy (XPS). It is equipped with a non–monochromatized Al K_α (1486.6 eV) and Mg K_α (1253.6 eV) X–ray radiation sources and a SPECS PHOIBOS–100 hemispherical analyzer. This study utilized Al K_α radiation at 100 W power.

XPS is an element sensitive method to determine the chemical composition of the surface [56]. It involves irradiation of a sample by photons of known energy (E=hv). As a consequence, electrons are liberated from their bound states with binding energy E_B and are analyzed with respect to their kinetic energy E_{KIN} (Fig. 3.4).

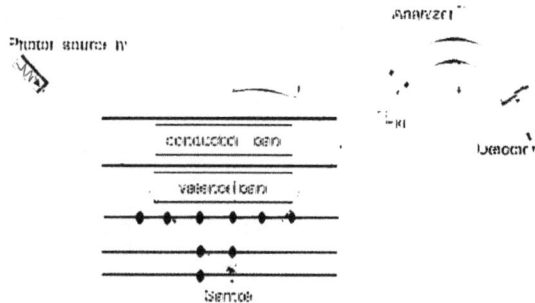

Fig. 3.4 X–Ray Photoelectron spectroscopy working principle, φ – the take–off angle under which the electrons are collected

The kinetic energy (E_{KIN}) of the emitted electrons is then given by the energy conservation law:

$$E_{KIN} = h\nu - \Phi - E_B \qquad (3.1)$$

where Φ stands for the analyzer work function, ν for the frequency and h is the Planck constant. The so–called photoelectric effect was originally postulated by Einstein [57].

The photoionization process can be illustrated in a simplified way as an initial and final state (Fig. 3.5):

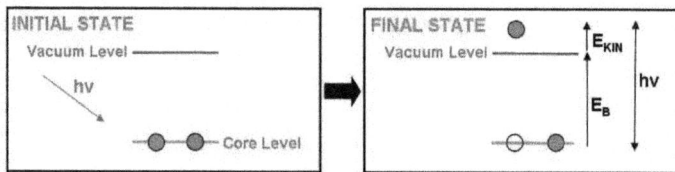

Fig.3.5 Photoelectron initial and final state picture

In the simplified one electron picture, the final state wave function represents an atom with a hole in the core level. If one assumes that the spatial distributions and energies of the final state electrons are the same as in the initial state before emission of the electron,

34

then the binding energy is simply equal to the negative orbital energy of the emitted electron. This approximation is called Koopmans' theorem and does not always correctly describe the photoionization effect.

The binding energy of an emitted electron including additional effects can be more correctly described by:

$$E_B = E_0 + \Delta E_{chem} + \Delta E_{ionic} + \Delta E_{relax} \tag{3.2}$$

E_0 stands for the undisturbed binding energy of the atom from which the photoelectron is excited and is tabulated in the literature [58].

$\Delta E_{chem.}$ is the chemical shift which is directly correlated with the change in electron density at the ionized atom when the atom is placed in a solid. Passing from the isolated atom to the compound layer, atomic orbitals become molecular orbitals. In consequence, valence electrons form molecular orbitals and core electrons change their energy due to the screening effects of the nucleus in the new chemical environment. Thus, E_B changes enabling chemical state identification of the element.

The influence of more distant atoms is given by ΔE_{ionic} which is also called "Madelung energy". This component is strongest in ionic compounds.

It is noted that the "sudden approximation" phenomena, which states that the primary excitation is rapid with respect to the relaxation process of the remaining electrons, provides a one–electron picture with the ability to explain the contributions to the binding energy mentioned so far.

The last term in the equation 3.2, ΔE_{relax} stands for complex multi-electron process corrections as i.e. shake–up and shake–off lines (e.g. excitation of plasmons) [59, 60].

In this work, the preparation of thin dielectric film was followed by quantitative analysis according to the formula:

$$\frac{N_A}{N_B} = \frac{I_A}{I_B} \cdot \frac{ASF_B}{ASF_A} \tag{3.3}$$

where N stands for the concentration of element A and B, I is the obtained XPS intensity and ASF stands for Atomic Sensitivity Factor. The ASF parameter was calibrated using additional Rutherford Backscattering Spectroscopy (RBS) measurements. More details are given in chapter 4.

3.2.1.2 X–ray Absorption Spectroscopy (XAS)

X–ray absorption spectroscopy (XAS) measurements are based on excitation of electrons from a core level to partially filled or even empty states in the same system [61]. Since the absorption peak positions and their intensity are directly related to the nature of unoccupied electronic states, XAS in contrast to XPS allows for probing of the empty orbitals.

XAS excitation obeys the dipole selection rule, which states that the change in angular momentum quantum number should be $\Delta l=\pm1$ between the initial and final states. The absorption of X–rays gives rise to an electronic excitation from a core level to an unoccupied state and a core hole is created.

Experimentally, XAS spectra are recorded by measuring either the electron yield or fluorescence yield as a function of incident photon energy. In this study, XAS experiments were carried out at the U49/2–PGM–2 undulator beamline of Bessy II using the ASAM end–station [62-64]. The XAS experiment requires a tunable source of energy, therefore the measurement must be done at synchrotron radiation facility.

Spectra were acquired in the total electron yield mode. To learn about the energetic positions of the lowest unoccupied states in the dielectric materials of interest, XAS O K-edge scans were performed. This measurement involved transitions from O1s to O2p (Fig. 3.6) states which are hybridized with the unoccupied d states of metal cations forming the conduction band minimum (CBM) [65]. As a result, the location of the CBM with respect to the Fermi level can be obtained [66]. In combination with UPS, which probes the occupied states at the valence band maximum (VBM), this method allows determination of the electronic band gap of the material under investigation [67].

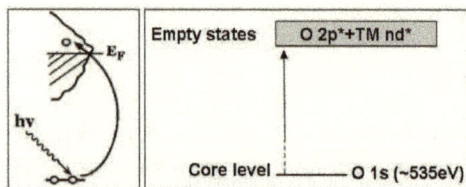

Fig. 3.6 Effects involved in the XAS emission (left) [69], schematic representation of the transition investigated by XAS at the OK_1 edge (right) [66].

3.2.1.3 X–Ray Reflectivity (XRR)

X–ray reflectivity (XRR) is a non–destructive method to study the thickness and roughness of thin layers and multilayer systems [70]. In this work, to measure XRR spectra, a Rigaku DMAX 1500 diffractometer of the Bragg–Brentano type and Rigaku Smart–Lab diffractometer with a CuK_α source (λ=1.5406 Å) were utilized.

In XRR, the sample is irradiated with a monochromatic X–ray beam of a known wavelength λ under an incident angle θ, and the reflected intensity is measured at an angle 2θ [71]. XRR operates typically in the θ range between 0 and 10 degrees. For an incident angle below the critical angle θ_C, a total external reflection is observed. Above the θ_C value, the X–ray penetrates into the sample. According to the electron density difference between the layer of interest and neighbouring films, oscillations (so called "Kiessig fringes") are observed. Due to the reflected beam interference, a certain periodicity and intensity fluctuation can be observed which gives the information about thickness and roughness of the film, respectively. Here, the thickness d is defined by the use of classical theory (Fresnel equation):

$$d \approx \frac{\lambda}{2\theta_{m+1} - 2\theta m} \tag{3.4}$$

where the denominator is the angular distance between two neighboring interference maxima $2\theta_{m+1}$ and $2\theta_m$ (see Fig. 3.7). To obtain the thickness, a model with electron density, surface roughness and thickness parameters is used. The simulation is done with the RCRefSim software [72].

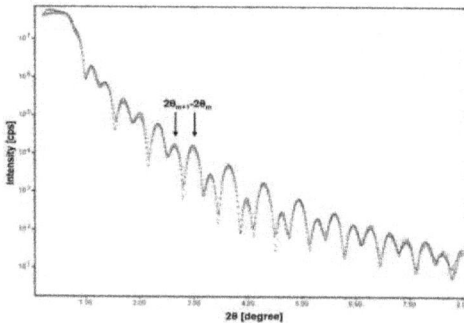

Fig. 3.7 Example of XRR measurement (red/print-dark grey) and simulation (green-light grey).

3.2.1.4 X–Ray Diffraction (XRD)

X–Ray Diffraction was performed using the same equipment as for XRR. The measurements were based on two different experimental geometries described below (Fig. 3.8): specular θ-2θ and grazing incidence XRD (GIXRD).

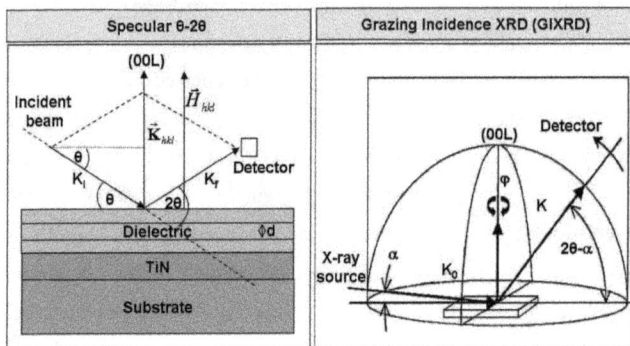

Fig. 3.8 XRD setup in two geometries: left θ–2θ specular scan, right–Grazing Incidence XRD (GIXRD).

Specular θ–2θ measurements

To obtain information about crystallographic structure of the film, specular θ–2θ measurements were performed. These scans involve higher θ angles than XRR ranging between 10° and 60° and reveal information about the crystallographic structure of the material.

In this process, electrons oscillate similar to Hertz dipole at a frequency of the incident beam and become the source of electromagnetic dipoles radiation. Due to the periodic arrangement of the atoms in a crystal and similar dimensions of the wavelength and interatomic distance, constructive or destructive interference appear, resulting in the observation of Bragg peaks.

Fig. 3.8 (left) shows the Laue condition in the vector form. The incident plane wave has a vector K_i whose length is $1/\lambda$. If no energy is gained or lost in the diffraction process (called elastic scattering) then the $|K_i|=|K_f|$, where K_f is the diffracted plane wave vector. A Bragg peak is observed when the impulse transfer vector K_{hkl} (equal to the difference

38

between K_i and K_f) is identical in the magnitude and direction to the reciprocal lattice vector H_{khl}.

The lattice d–spacing in a single crystal has the following relation to the reciprocal lattice vector [73]:

$$d_{hkl} = \frac{1}{|H_{hkl}|} \tag{3.5}$$

Using the geometrical relationship shown in Fig. 3.8 (left),

$$\sin\theta = \frac{|K_{hkl}|}{2} \cdot \frac{1}{|K_i|} \tag{3.6}$$

allows deriving Bragg's law for X–ray diffraction from the Laue condition in vector form:

$$\frac{1}{d_{hkl}} = 2K_i \sin\theta = 2\frac{1}{\lambda}\sin\theta \tag{3.7}$$

As a result, the scattering of X–rays on crystal planes with a spacing d is described by:

$$n \cdot \lambda = 2 \cdot d_{hkl} \cdot \sin\theta \tag{3.8}$$

where n is the order of reflection (integer) and θ is the Bragg reflection angle. For amorphous materials with random atomic orientation, no diffraction peaks are observed. If atoms are arranged in a periodic form, the waves diffracted from the parallel atomic planes can interfere constructively resulting in strong diffraction maxima characteristic for a given material according to Bragg's law (Eq. 3.8).

Grazing Incidence XRD (GIXRD) Analysis

To gain surface sensitive crystallographic information about thin dielectrics used in this work, a so called Grazing Incidence XRD (GIXRD) study was performed. Grazing–incidence diffraction is a scattering geometry combining the above described Bragg condition with the conditions for X–ray total external reflection from crystal surfaces. Due to the strong enhancement of the electric field at the surface under total reflection condition, this provides superior sensitivity of GIXRD as compared to the other diffraction schemes in the studies of thin surface layers, since the penetration depth of X–rays inside the slab is reduced by three orders of magnitude–typically from 1–10 μm to

1–10 nm [74]. In our experiment, the X–ray beam hits the surface at grazing incidence ($\alpha=0.8°$) while the detector angle is scanned over a 2θ range from $24°$ to $55°$.

3.2.1.5 Atomic Force Microscopy (AFM) for roughness determination

The AFM was developed to overcome a basic drawback with scanning tunneling microscopy (STM) [75] –which could image topology of conducting or semiconducting surfaces only. The AFM has the advantage of imaging almost any type of surface. Most AFMs use a laser beam deflection system (shown in Fig. 3.9), introduced by Meyer and Amer [76]. In this work, we used an Omicron variable temperature (VT) UHV AFM. It operates by measuring attractive or repulsive forces between a tip and the sample [77]. In the repulsive mode, called "contact mode", the tip placed at the end of a spring cantilever touches the surface. Two scanning modes may be distinguished in this case: constant height mode and constant force mode. The former one is described elsewhere [78]. This study utilized constant force mode. In this case, the cantilever vertical deflection reflects repulsive force acting upon the tip. The deflection is maintained by the feedback circuitry on the preset value. With changing topology the tip–sample force is kept constant while the piezotube is moving. This allows Z–coordinate (height) acquisition. For X and Y information, light deflection from the back side of the tip is used and monitored by a Position Sensitive Detector (PSD). Main advantage of the constant force mode is the possibility to measure with high resolution simultaneously topography and other characteristics, e.g. friction forces, spreading resistance or leakage current.

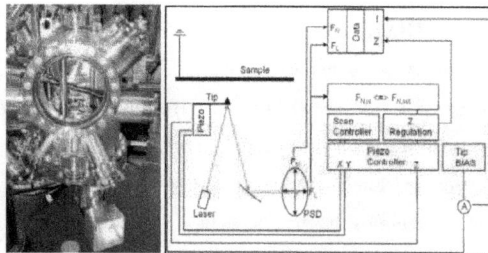

Fig. 3.9 Left: Omicron UHV Atomic Force Microscope (UHV C–AFM); Right: Schematical AFM working principle

3.2.2 Dielectric and electrical characterization

3.2.2.1 Capacitance–voltage (C–V)

To measure capacitance of the prepared MIM capacitors, C–V measurements were performed. In this measurement, a small AC signal (dV_{AC}~20 mV) is superimposed on a DC (V_{DC}) signal and applied to the plates of the capacitor (see Fig. 3.10, left). Subsequently the DC voltage is varied in a desired range (in this study: –1 V÷1 V) and a relationship between the capacitance and the DC bias voltage is obtained. CV measurements were performed using Keithley Semiconductor Parameter Analyzer 4200-SCS with integrated auto balancing bridge technique [79]. In this measurement method, the impedance |Z| of the capacitor and the phase shift φ on the capacitor are measured. Subsequently, the capacitance value is calculated assuming an appropriate equivalent circuit which best describes the capacitor [80]. In this study, the tested capacitors were modeled as non–ideal elements composed of a parallel combination of a capacitor and a resistor (Fig. 3.10, right). According to this model, the capacitance can be expressed as:

$$C_p = \frac{1}{|Z|\omega\sqrt{1+\dfrac{1}{Q^2}}}$$

(3.9)

where ω is the angular frequency and Q stands for quality factor and is expressed as:

$$Q = |\tan\varphi|$$

(3.10)

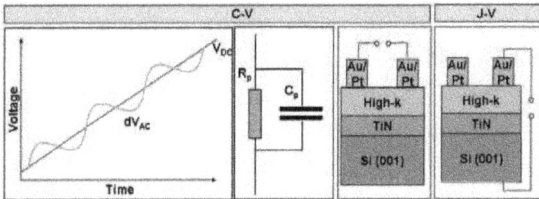

Fig. 3.10 C-V: Applied AC amplitude followed by DC voltage sweeping the bias range (left), parallel equivalent circuit for C–V acquisition (middle), simplified schematics of C–V measurement; J-V: Simplified drawing of J–V measurement setup.

The capacitance values were then used to extract the dielectric constant of the investigated dielectrics according to the parallel plate capacitor model where C is described in Eq. 2.6.

The simplified schematic of C–V measurement setup is shown in the Fig. 3.10.

3.2.2.2 Current–voltage (J–V)

Macroscopic leakage current curves (J–V) were acquired based on the configuration shown in Fig. 3.10. The characteristics can be recorded by using a simple measurement setup consisting of a DC voltage source and a precise current meter [81]. The DC leakage current was obtained at room temperature while sweeping the applied bias voltage. For leakage current mechanism study, temperature dependent J–V characteristics were acquired in the temperature range of 210–390 K using SUSS PMV–200 Vacuum Prober.

3.2.2.3 Conductive Atomic Force Microscopy (C–AFM)

Nanoscopical leakage current investigation was carried out using a C–AFM instrument. It utilizes the same working principle as the above described AFM. Additionally to the topography measured in contact mode, leakage current information is acquired [82, 83]. The experiments were done under UHV conditions (10^{-11} mbar) and prior to the C–AFM studies the samples were heated indirectly (radiative) at 150 °C in order to free the film from hydrocarbon and water contamination. The conditions in UHV chamber (residual gases' desorption) were controlled using an Inficon mass spectrometer. We used diamond coated tips due to the wear off resistance and low resistivity with a force constant of 2.8 N/m. The sample was grounded and the tip polarized. Current maps were acquired at a constant voltage. At certain points of interest, J–V characteristics were measured. The acquisition was followed by data evaluation using SPIP Software [84].

Chapter 4

Results and discussion

This chapter summarizes the experimental results. It is divided into two parts: in the first part the focus is put on comparison of $BaHfO_3$ and HfO_2 dielectric layers on TiN metallic substrate; second part shows an approach to optimize the dielectric properties of thin $BaHfO_3$ films by introducing Ti ions on the B position of the cubic ABO_3 perovskite structure.

4.1 Characteristics of BaHfO₃ dielectric films

4.1.1 Macroscopic study

This subchapter is split into 4 parts with regard to the goal of the analysis: chemical composition, structural properties, electrical characteristics, and band gap investigation.

4.1.1.1 Chemical composition

To achieve the target stoichiometry, BaO and HfO_2 fluxes were calibrated using quantitative analysis of XPS spectra. In this analysis, if a photon of energy $h\upsilon$ ionizes a core level X in an atom of element A in a solid, the photoelectron current I_A is given by:

$$I_A(X) = K\sigma_A(h\upsilon, X)\beta_A(h\upsilon, X)\overline{N}_A\lambda_M(E_A)\cos\theta \qquad (4.1)$$

where $\sigma_A(h\upsilon, X)$ is the photoelectric cross–section for ionization of the X core level by photons with energy $h\upsilon$, $\beta_A(h\upsilon, X)$ the angular asymmetry parameter for above emission, $\overline{N}_A$ the atomic density for the element A averaged over the analysis depth, λ_M the inelastic mean free path (IMFP) in matrix M containing atom A at kinetic energy E_A ($E_A = h\upsilon - F_B$), and θ is the angle of the emission to the surface normal. K is a proportionality constant containing fixed operating conditions (X–ray characteristic line flux (J_0), transmission ($T(x, y, E_A)$) and efficiency $D(E_A)$ of the analyzer at a given E_A).

43

Values for σ_A and β_A have been calculated by Scofield and Reilman, respectively [85, 86]. For a homogenous material, the photoemission line intensity can be expressed as:

$$I_A \propto C_A \sigma_A(h\upsilon) N_A \lambda_M(E_A) \qquad (4.2)$$

where C_A is a constant containing the proportionality constant K, angular asymmetry and angle of the emission. The formula for $\lambda_M(E_A)$ was derived by Seah and Dench [87]:

$$\lambda_M(E_A) = A \cdot (d \cdot E_A)^{1/2} \quad \text{where} \quad A = 0.41 nm^{1/2} eV^{-1/2} \qquad (4.3)$$

where $\lambda_M(E)$ is in monolayers, d is the "monolayer" thickness in nanometers ($d^3 = \dfrac{A}{\rho \cdot n \cdot N} x 10^{24}$ with A-atomic or molecular weight, ρ-bulk density in kg·m^{-3}, n– number of atoms in molecule and N–Avogadro number) and E is in electron volts. If we define an atomic sensitivity factor for an element A as ASF_A:

$$ASF_A = C_A \sigma_A(h\upsilon) \lambda_M(E_A) \qquad (4.4)$$

then the intensity from Eq. 4.2 is given by:

$$I_A = N_A\ ASF_A \qquad (4.5)$$

If we wish to know the relative material amounts in a compound formed from material A and B, it is necessary to know the sensitivity factors for the elements and to measure the area under XPS photoemission lines [88, 89]:

$$\frac{N_A}{N_B} = \frac{I_A / ASF_A}{I_B / ASF_B} \qquad (4.6)$$

In this work, the ASF parameters were calibrated using independent analysis of chemical composition based on Rutherford Backscattering Spectroscopy (RBS) measurement. The correction was below 5 % with respect to the single elements data given in XPS handbooks [60].

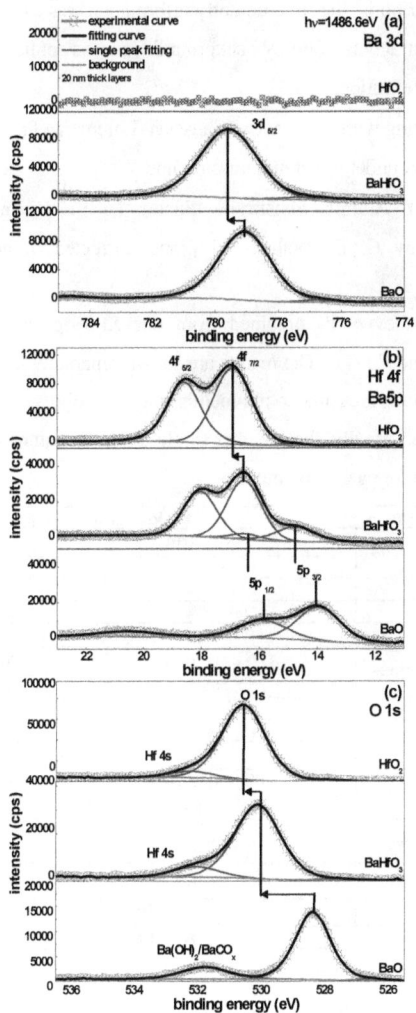

Figure 4.1: XPS for HfO$_2$, BaHfO$_3$ and BaO compounds; (a) Ba 3d (b) Hf 4f/Ba5p, (c) O 1s photoemission spectra. The red (print light grey) plot presents the raw data; the black line shows the fitted curve; single peaks are marked in blue (print dark grey) and grey background represents the Shirley function. All layers with similar 20 nm thickness.

According to this description, the stoichiometry information was calculated in five steps:

1. Background extraction – Shirley background type was applied,
2. Energy range definition,
3. Single peak fitting with asymmetric Gaussian–Lorentzian function,
4. Area integration under the photoemission line,
5. Application of Atomic Sensitivity Factor (ASF) obtained from [58] as demonstrated by C. D. Wagner [90] and corrected in accordance to RBS measurements.

Table 1 summarizes the results obtained from our XPS spectra analysis (Fig. 4.1) according to Eq. 4.5 and Eq. 4.6. The results are shown separately for each line (Ba $3d_{5/2}$, Hf 4f, O 1s) and contain area under photoemission line followed by data divided by respective ASF value. Finally, below the single lines, a summary result from the measurement is shown for each compound.

Dielectric compound	Ba $3d_{5/2}$ (ASF=7.469)	
	XPS peak area	Area/ASF
HfO_2	15 (noise)	2.0 (noise)
$BaHfO_3$	219900.5	29441.8
BaO	204718.9	27409.14
	Hf 4f (ASF=2.639)	
	XPS peak area	Area/ASF
HfO_2	305568.5	115789.5
$BaHfO_3$	92271.3	31964.5
BaO	15 (noise)	5.7 (noise)
	O 1s (ASF=0.711)	
	XPS peak area	Area/ASF
HfO_2	162032.63	227894.9
$BaHfO_3$	65575	92230
BaO	19566.7	27520
	FINAL RESULT (including ASF)	
HfO_2	Hf:O= 1:1.97	
$BaHfO_3$	Ba:Hf:O = 1:1.08:3.13	
BaO	Ba:O = 1:0.992	

Table 1: Example of quantitative analysis for HfO_2, $BaHfO_3$ and BaO obtained from XPS measurement. In the left column, the integrated area under XPS curve, right column includes the atomic sensitivity factor in the calculation. For final result, the right column is considered

All samples taken into further consideration exhibited the target BaHfO₃ stoichiometry (Ba: Hf: O~1:1:3) with a maximal deviation of 5 %.

The analysis of chemical shifts of the Ba $3d_{5/2}$, Hf 4f and O 1s photoemission lines is illustrated in Fig. 4.1. Here, the sample thicknesses were close to 20 nm. The discussion starts from the marginal values of the pure compounds (here HfO₂ and BaO), then they are compared with BaHfO₃ compound with regard to the chemical shift.

Fig. 4.1 (a) shows Ba 3d photoelectron emission in the energy range from 774 to 785 eV. For pure HfO₂ film, only noise signal is observed as expected. In BaO thin layer, the Ba $3d_{5/2}$ peak appears on the reference position 7 79 eV. This is in agreement with literature data for oxidized Ba²⁺ valence states [91]. The photoemission line of the intermediate compound, BaHfO₃, appears on higher binding energy (779.6 eV). The shift in binding energy equals 0.5 eV and is caused, according to the electronegativity rules, by an increased ionicity of Ba in BaHfO₃ with respect to BaO [92].

Fig. 4.1 (b) shows the photoemission lines in the energy window from 10 to 23 eV. For HfO₂, two peaks centered at 16.5 eV and 18.2 eV are clearly visible. These signals are characteristic of Hf $4f_{7/2}$ and Hf $4f_{5/2}$ photoemission lines in fully oxidized HfO₂ compounds. The binding energy distance of spin–orbit–splitting components (SOS) is close to the handbook XPS data [58] and equals 1.7 eV. For BaO, the photoemission energy range detects the Ba 5p line splitting into $5p_{3/2}$ (14.0 eV) and $5p_{1/2}$ (15.5 eV) [93, 94]. No contribution from Hf component is observed. For BaHfO₃ two main peaks can be clearly identified: the Hf $4f_{7/2}$ and Hf $4f_{5/2}$ components distanced by 1.7 eV. The binding energy shift with regard to HfO₂ is equal to $\Delta E = -0.4$ eV. In the initial state picture this can be explained by higher electronegativity of Hf compared to Ba ions (1.3 and 0.89 in Pauling scale, respectively) [95]. A small shoulder on the lower binding energy (~13 eV) can be attributed to Ba 5p states which overlap with the Hf 4f line. Here, the Ba 5p components are also visible; however the intensity is almost two times lower.

The corresponding O 1s spectra are shown in Fig. 4.1 (c). For HfO₂ the main O 1s peak is located at 530.9 eV. For BaO it is visible at 528.7 eV. The BaHfO₃ exhibits an O 1s photoemission line between HfO₂ and BaO at 530.4 eV. According to the electronegativity rules, this indicates the formation of a mixed BaO–HfO compound [32]. Furthermore, in case of inhomogeneous mixed compounds (e.g. phase separated HfO$_x$

and BaO_x), two photoemission peaks would be visible due to sufficient XPS energy resolution. In this study, however, for $BaHfO_3$, only one peak in the O 1s energy range was resolved. This indicates the formation of a homogenous $BaHfO_3$ film. At higher binding energy in HfO_2 and $BaHfO_3$, the Hf 4s peak appears (532.4 eV). In BaO the additional peak around 532 eV is attributed to the formation of $Ba(OH)_2$ and/or $BaCO_x$ [96] at the film surface.

Summarizing results from this subchapter, simultaneous BaO and HfO_2 deposition results in the formation of a homogenous mixed oxide layer with the stoichiometry of $BaHfO_3$ (within 5 % analytical error). This result is further corroborated by the following XRD structure investigation.

4.1.1.2 Structural properties

The deposition and in–situ chemical characterization process was followed by ex–situ investigations. This part summarizes the experimental results obtained by XRD. The data present the structural properties of HfO_2 and $BaHfO_3$ layers.

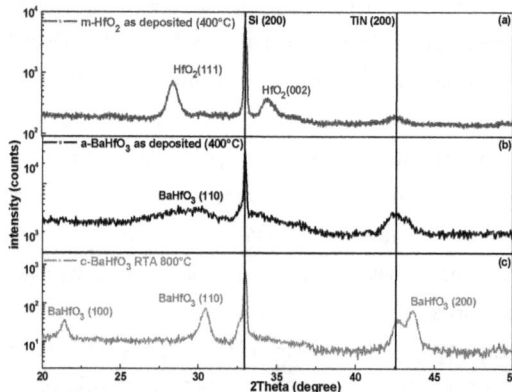

Figure 4.2: Specular θ–2θ analysis from (a) as deposited 21 nm monoclinic HfO_2, (b) as-deposited 43.5 nm amorphous $BaHfO_3$ and (c) 43.3 nm cubic $BaHfO_3$ after post-deposition treatment

48

Fig. 4.2 (a) shows a specular θ–2θ scan obtained for a 21 nm thick HfO_2 film. The as deposited (400 °C) layer shows diffraction peaks corresponding to the monoclinic HfO_2 phase (m-HfO_2) [35]. Additionally, the (200) diffraction peak from the TiN metal electrode as well as the (200) peak from Si substrate are visible. The latter one is normally forbidden in kinematical scattering theory but it can be visible due to the covalent bonding character of Si which results in a deviation from spherical electron distribution around Si atom. In addition, further intensity change might result on the Si (200) Bragg peak position from so called "Umweganregungen" according to the experimental geometry [97].

Fig. 4.2 (b) shows the diffraction pattern from as deposited 43.5 nm thick $BaHfO_3$ layer. After the deposition process, no reflection apart from the Si and TiN signals are visible. Clearly, stoichiometric admixtures of Ba atoms into the pure HfO_2 layers prevent the film to crystallize and the compound is X-ray amorphous (a–$BaHfO_3$). The XRD pattern changes dramatically after RTA treatment at 800 °C. For the a–$BaHfO_3$ layers, post–deposition thermal treatment induces crystallization and the XRD spectrum is a finger print of the cubic $BaHfO_3$ perovskite structure [98] with lattice constant a= 4.16 Å, as will be discussed later in the subchapter 4.2.1.2. It is noted that for thin $BaHfO_3$ layers below 10 nm which are the goal for DRAM application, the crystallization process starts at higher temperature and the samples remain amorphous even after RTA at 800 °C (data not shown). In this case, higher annealing temperatures are required (~900 °C) to induce crystallization towards the c–$BaHfO_3$ phase.

4.1.1.3 Electrical characteristics

We now present the results of electrical characterization for Pt/high–k/TiN MIM structures. Fig. 4.3 (a) compares the dielectric constants of the investigated insulator layers. Here, CET (introduced in Eq. 2.8 and Eq. 2.11) was plotted as a function of physical thickness. The k–values were extracted according to the slope of the linear regression of single points obtained from C–V measurements. The intersection with zero thickness point gives information about the existence of low-k interface.

The k–value of the monoclinic HfO_2 is around 19, which is in good agreement with literature [36].

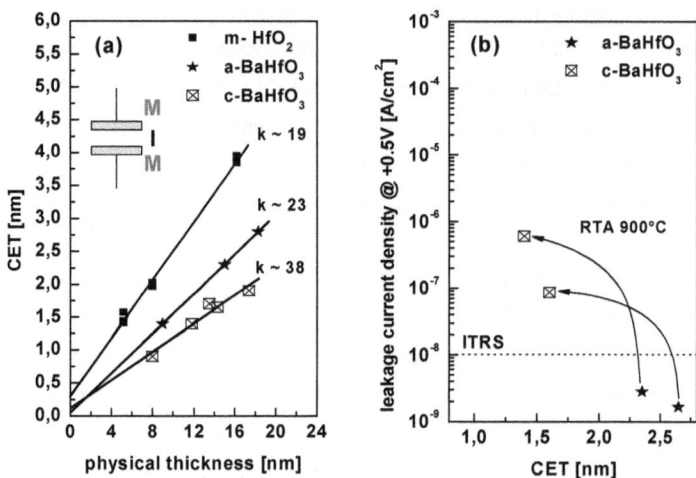

Figure 4.3: (a) Comparison of the dielectric constant values for m–HfO$_2$, a–BaHfO$_3$ and c-BaHfO$_3$; (b) Influence of the RTA process on the leakage currents in BaHfO$_3$ dielectric.

Admixture of Ba results in a slight increase of the dielectric permittivity so that a value of about k~23 is obtained for amorphous BaHfO$_3$ compound. Crystallization in the c-BaHfO$_3$ phase upon thermal treatment (proved by XRD) is accompanied by an abrupt increase of the dielectric constant to about 38. Furthermore, for c–BaHfO$_3$, the presence of an interface layer within the ±0.1 nm error can be excluded from the plot. The CET values for m–HfO$_2$ are twice higher than for c–BaHfO$_3$: for the same thickness (8 nm) the CET for m–HfO$_2$ is close to 2 nm whereas the same c–BaHfO$_3$ thickness results in CET close to 0.9 nm.

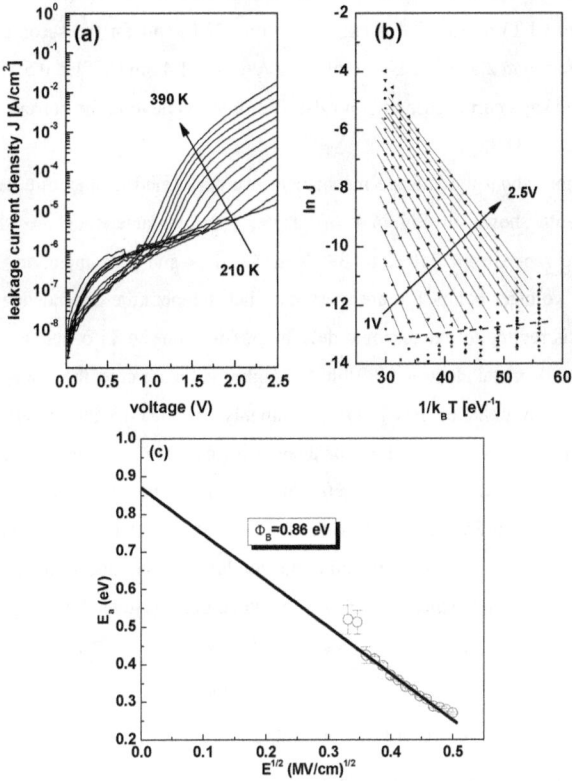

Figure 4.4: (a) Temperature dependent J–V characteristics measured for 8 nm thick c-BaHfO$_3$ layer; (b) Corresponding Arrhenius plot and (c) Activation energy (E$_a$) as a function of square root of electric field (E$^{1/2}$)

Figure 4.3 (b) illustrates the influence of the RTA treatment on leakage currents in BaHfO$_3$ films. Two samples were selected to show the J–V behavior of this dielectric. As can be seen from Fig. 4.3 (b), amorphous BaHfO$_3$ layers exhibit leakage current values of 10^{-8} A/cm^2. RTA induced crystallization of these films results in a significant reduction of the CET (<1.5 nm) which is a result of the improved dielectric constant. As a consequence, a substantial increase in the leakage current density (from $1.7 \cdot 10^{-9}$ A/cm^2 (@ 2.65 nm CET) to $8.6 \cdot 10^{-8}$ A/cm^2 (@1.6 nm CET) and for the second sample from $2.8 \cdot 10^{-9}$ A/cm^2 (@ 2.35 nm CET) to $6 \cdot 10^{-7}$ A/cm^2 (@ 1.4 nm CET) at 0.5 V) is observed. Thus, the leakage current density on c-BaHfO$_3$ layer is beyond the current limit (dashed line) defined by ITRS.

To learn more about the leakage mechanism, we performed temperature dependent J–V measurements shown in Fig. 4.4 (a). Here, J–V characteristics are shown in the temperature range from 210 K to 390 K under TiN substrate injection. The currents detected at voltages below 1 V are noisy and their temperature dependence is weak. We attribute this regime to macroscopic defects, possibly due to Ti contamination from the underlying TiN metal electrode [99]. In the higher voltage range, the leakage mechanism is dominated by point defects [100], presumably associated with so called "hot spot" areas. In this strongly temperature dependent voltage regime, the concentration of carriers is dominated by trapping and detrapping phenomena, with voltage dependence resembling a Poole–Frenkel effect. In the Pool–Frenkel model, field–enhanced electrons trapped in localized states are expelled to the conduction band due to thermal fluctuations. The electric field dependence of the Pool–Frenkel current is given by [101]:

$$J_{PF} = A \cdot E \cdot \exp\left[-\frac{q}{k_B T} \cdot \left(\phi_B - \sqrt{\frac{qE}{\pi \varepsilon_\infty \varepsilon_0}} \right) \right] \qquad (4.7)$$

where A is the effective Richardson constant, E is the electric field, q–elementary charge, k_B–Boltzmann constant, T–temperature, ϕ_B –effective trap depth, and ε_∞ –the high–frequency dielectric constant (equal to the squared optical refraction index)[1], and ε_0 the dielectric permittivity of free space. Here, E is the electric field given by the applied voltage V and film thickness d:

$$E = \frac{V}{d} \qquad (4.8)$$

---

[1] Note that the dielectric constant in Eq. 4.7 is the high–frequency ε_∞, not the low–frequency ε_r. This is because the atomic vibrations which are responsible for ε_r are not fast enough to respond to the motion of an electron [28].

The determination of the effective trap depth ϕ_B from J–V characteristic is as follows: the current–temperature dependence of Eq. 4.7 can be rewritten as:

$$\ln(J_{FP}) = \ln(A) + \ln(E) - \frac{1}{k_B T} \cdot q\left(\phi_B - \sqrt{\frac{qE}{\pi \varepsilon_\infty \varepsilon_0}}\right) \qquad (4.9)$$

$$\ln(J_{FP}) = \ln(A) + \ln(E) - \frac{1}{k_B T} \cdot q\left(\phi_B - \beta\sqrt{E}\right) \text{ with } \beta = \sqrt{\frac{q}{\pi \varepsilon_\infty \varepsilon_0}} \qquad (4.10)$$

Further discussion focuses on extraction of ϕ_B and β parameter from the electrical data obtained by J–V. The Pool–Frenkel effect is schematically shown in the Fig. 4.5. Here, the solid lines show the Coulomb barrier without externally applied electric field, the dashed line shows how the potential profile changes after the application of electric field. The term ϕ_B from Eq. 4.7 is marked on the plot.

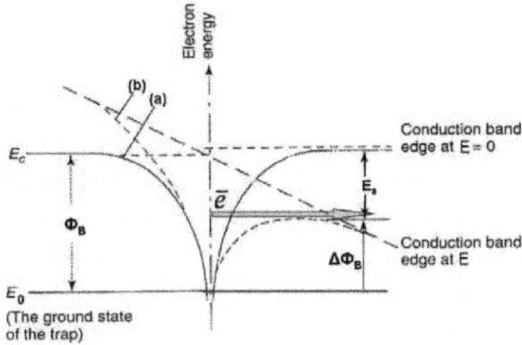

Figure 4.5: Schematical representation of Pool–Frenkel effect [102], (a)–trap without applied voltage, (b) trap with applied voltage

In this study, we can use the data obtained from J–V acquisition and extract the ϕ_B value. The graphical representation of Eq. 4.10 known as Arrhenius plot is shown in Fig. 4.4 (b). Each group of symbols on the plot, corresponds to a certain value of applied voltage (ranging from $1\div2.5$ V) measured at different temperatures (210–390 K). As seen from Eq. 4.10, the slope varying indicates the field dependence of the activation energy E_a:

$$E_a = \phi_B - \beta\sqrt{E} \qquad (4.11)$$

The dependence of the activation energy on the square root of the electric field is shown in Fig. 4.4 (c). The trap depth ϕ_B is obtained from the intercept with the y-axis (vanishing applied voltage). In the case of BaHfO$_3$ discussed here, a trap depth of 0.86 eV was extracted.

As HfO$_2$–based materials are reported to be an n–type material, the trap energy level can be interpreted as an electron trap level below the conduction band [52, 103]. Moreover, the curve slope β can be determined and compared with theoretical value expected for Pool–Frenkel emission. In our case, it equals to $3.6E^{-5}$ $(m \cdot V)^{1/2}$ (extracted from the slope in Fig. 4.4 (c)). We can conclude that the plot resembles the ideal Pool-Frenkel behavior, as the ß value is close to the theoretical value of $3.8E–5$ $(m \cdot V)^{1/2}$ within a deviation of 5%. This confirms the Pool-Frenkel effect as a dominating mechanism of leakage current for the discussed range of applied voltages in case of BaHfO$_3$ films on TiN metal electrodes.

In summary, as a result of Ba incorporation into HfO$_2$, the dielectric permittivity of the amorphous dielectric increased by about 20% from 19 to 23 for materials deposited at 400 °C. Thermal treatment of BaHfO$_3$ (RTA~800 °C–900 °C) resulted in a crystallization in the c–BaHfO$_3$ phase and the k–value increased to ~40. As a result of dielectric constant increase, the CET values were scaled down to <1 nm. However, an abrupt increase in leakage current density was observed ($6 \cdot 10^{-7}$ A/cm^2 at 0.5 V for CET=1.4 nm). Temperature dependent J-V analysis allows concluding that the leakage current is governed by the Pool–Frenkel effect in case of c-BaHfO$_3$ films on TiN metal electrodes.

4.1.1.4 Band gap and band alignment

This subchapter addresses the fundamental electronic structure of HfO$_2$ and BaHfO$_3$ on TiN. To explain the origin of leakage current and gain insight into the band alignment, we performed a combined XPS–XAS study. In this method, the occupied states are probed by synchrotron radiation excited valence band photoemission, while the unoccupied states are sampled by XAS measurements.

54

Figure 4.6 compares the VB photoemission spectra (a) and the XAS spectra at the O K-edge (b) for the HfO_2 and $BaHfO_3$ films. The valence band maxima (VBM) for the investigated materials shown in the Fig. 4.6 (a) are formed by O 2p states. For HfO_2, the VBM is located at around 3.7 eV below the TiN Fermi level. The corresponding values for a–$BaHfO_3$ and c-$BaHfO_3$ are 3.6 and 3.7 eV, respectively. It is therefore concluded, that the VB offsets for both a–$BaHfO_3$ and c–$BaHfO_3$ layers on TiN do not differ significantly with respect to the values obtained for m–HfO_2.

To investigate the energetic position of the lowest unoccupied levels, XAS O K–edge scans were performed (shown in Fig. 4.6 (b)). Here, by applying dipole selection rules as described in 3.2.1.2, the unoccupied part of the O 2p final states, which hybridize with Hf and Ba metal d states, can be reached from the initial O 1s core level. [104, 105]. It was shown that the conduction band edge (CBE) is to a very good approximation defined by the position of the leading XAS peak [67]. The location of the CBE relative to the Fermi level is obtained using the binding energy of the O 1s peak in the corresponding XPS spectrum [106]. For HfO_2 film, the O K–edge reflects the unoccupied Hf 5d states which hybridize with O 2p valence states. The leading absorption peak is located at around 532.9 eV. For a–$BaHfO_3$ and c–$BaHfO_3$ the O K–edge shows the Ba 5d states located at 532.6 and 532.5 eV, respectively.

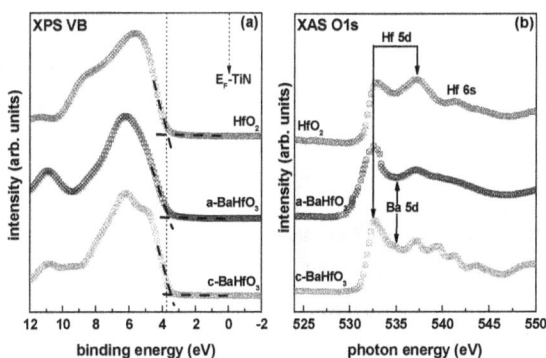

Figure 4.6: (a) Valence band spectra of HfO_2, a–$BaHfO_3$ and c–$BaHfO_3$ acquired at hυ=200 eV; (b) corresponding XAS spectra at the O K–edge.

From the comparison of XAS and XPS peak positions, the band alignment was deduced as shown in Table 2. Below the table, a graphical representation of the data is shown.

	VBO	XAS leading peak position	XPS (O 1s) main line	CBO=XAS–XPS	BG=VBO+CBO (±0.1 eV)
HfO$_2$	3.7	532.9	530.9	2.0	5.7
a–BaHfO$_3$	3.6	532.6	530.4	2.2	5.8
c–BaHfO$_3$	3.7	532.5	530.2	2.3	6.0

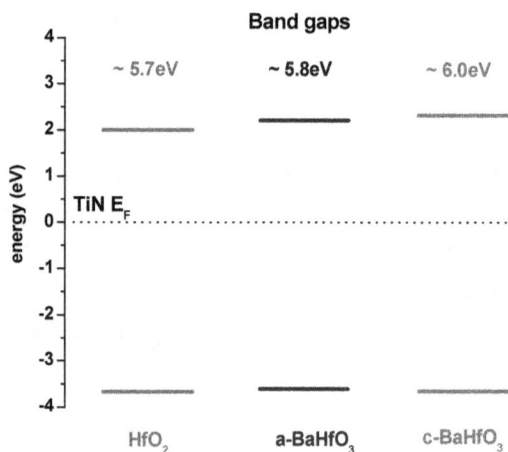

Table 2: Band offsets and band gap calculation based on UPS, XAS, and XPS; notation: VBO–valence band offset, CBO–conduction band offset, BG–band gap; below the table, a graphical representation of band gaps and band offsets is shown

Summarizing, Ba addition to HfO$_2$ results in nearly no shift in the VBO, even after application of the RTA treatment for BaHfO$_3$ (around 3.7 eV in Table 2). A slightly different situation is observed in the CBO region. After Ba ions incorporation in HfO$_2$, the CBO shifts by about $\Delta E=0.3$ eV towards higher values. In consequence, the formation of amorphous $BaHfO_3$ does not result in a significant change of the BG with respect to HfO_2. After RTA, BG remains within the experimental error nearly unchanged with a value of ~6.0 eV for $c-BaHfO_3$.

4.1.2 Nanoscopic investigation

The macroscopic study of leakage current mechanism in $BaHfO_3$ layers presented in the subchapter 4.1.1 is now supplemented by nanoscopic investigation to reveal spatial inhomogeneities in the leakage current characteristics.

4.1.2.1 Conductive Atomic Force Microscopy (C–AFM)

Here, we focus on the leakage current behavior of HfO_2 and $BaHfO_3$ samples investigated on the nanoscale using C–AFM. For this study, we chose only crystalline layers ($m-HfO_2$ and $c-BaHfO_3$). Leakage currents of the corresponding $a-BaHfO_3$ layers were below the detection limit of our C–AFM equipment (10 pA).

Figure 4.7 shows data obtained on 5 nm $m-HfO_2$ layer. The scan area was 500x500 nm^2. The film was measured at two voltages (2 V in Fig. 4.7 (b) and 4 V in Fig. 4.7 (c)) and as a reference, a 0 V measurement is shown (Fig. 4.7 (a)). First we discuss the topology, and than the current maps. The results from this investigation are summarized in Table 3. Both topology and current are followed by respective histograms. The Gaussian distribution shows random feature appearance whereas plot skewness introduces values higher or lower than the average value. This suggests higher or lower current trend, respectively.

The bright spots on the topology pictures (Fig. 4.7 (a)–(c)) refer to high topology values. The dark points mark lower positions in the topology picture.

In the topology picture, we obtain maximum height values (Z_{range}) of about 1.6 nm and this is similar for all acquisitions. Voltage application to the layer (2 V in Fig. 4.7 (b) and 4 V (c)) does not lead to a significant change in the topology: the information is obtained by the root mean square (RMS) value which is for scans at 0 V, 2 V and 4 V equal to 1.02 nm, 1.013 nm and 0.98 nm, respectively. Additionally, it is clear from the histograms shown in Fig. 4.7 (d) that for all three acquisitions the curves resemble a

Gaussian shape. This suggests a random grain distribution characteristic for polycrystalline samples [107]. The main conclusion here is the stability of the sample surface under voltage scans, suggesting that the film is not modified and the electrical data are comparable.

As a next step, the leakage current measured simultaneously with the topology data is described (Fig. 4.7 (e)–(g)). Here, the white positions refer to the low resistive points of the measured area. The dark areas represent higher insulating regions. In the leakage current map taken at 0 V (Fig. 4.7 (e)) only a noise signal is visible. The maximal current value (J_{range}) of around 14 pA and the RMS value for leakage current RMS_{leak}=6.7 pA are obtained. With the preamplifier used in this study, this is a very stable and low noise result. After voltage application (2 V in Fig. 4.7 (f)), the map becomes non–uniform: single "hot spots" with a maximum value of ~716 pA are visible and the RMS_{leak}=132.5 pA is demonstrated. The amount of bright spots and non–uniformity of current map increases after further voltage increase (4 V in Fig. 4.7 (g)): the leakage magnitude intensifies with a maximum J_{max}=2.38 nA and RMS_{leak}=956.6 pA. The histograms below the current maps (Fig. 4.7 (h)) show the current distribution for all three cases: the reference measurement exhibits Gaussian distribution which is typical for the noise signal and random current values; the map acquired at 2 V shows the tendency for hot spot establishment with currents higher than the average value (positive skewness of the histogram); for the measurement at 4 V most values are gathered around 1 nA, suggesting a different current behavior. In the latter case, the whole layer becomes more leaky with single weak insulating points. Here, the background is insulating and the leakage path is dominated by the mentioned "hot spots".

Now we compare the topology maps and parallel current acquisition. As the 0 V measurement exhibits current noise level of the C-AFM, it is not further discussed. For the measurement at 2 V the correlation between topology and current maps is not straightforward. In comparison, the weak points on the current map at 4 V correlate to lower positions in the topology and are therefore not attributed to the material properties but to film thickness inhomogeneities due to the deposition technique.

Figure 4.8 shows four correlation examples between topology and leakage current map. It illustrates that low topology regions exhibit high leakage behavior. This result indicates that a more homogenous film deposition technique like CVD results in more insulating HfO_2 layers.

HfO_2	Z_{range}	RMS_{topo}	J_{range}	RMS_{leak}
0 V	1.61 nm	1.02 nm	14.01 pA	6.7 pA
2 V	1.68 nm	1.013 nm	715.7 pA	132.5 pA
4 V	1.54 nm	0.98 nm	2.38 nA	956.6 pA

Table 3: C–AFM acquisition on 5 nm HfO_2 layer at 0 V, 2 V and 4 V; topological (Z_{range}-maximum topology value, RMS_{topo}-root mean square from topology picture) and electrical (J_{range}-maximum current, RMS_{leak}-root mean square values for current maps) results

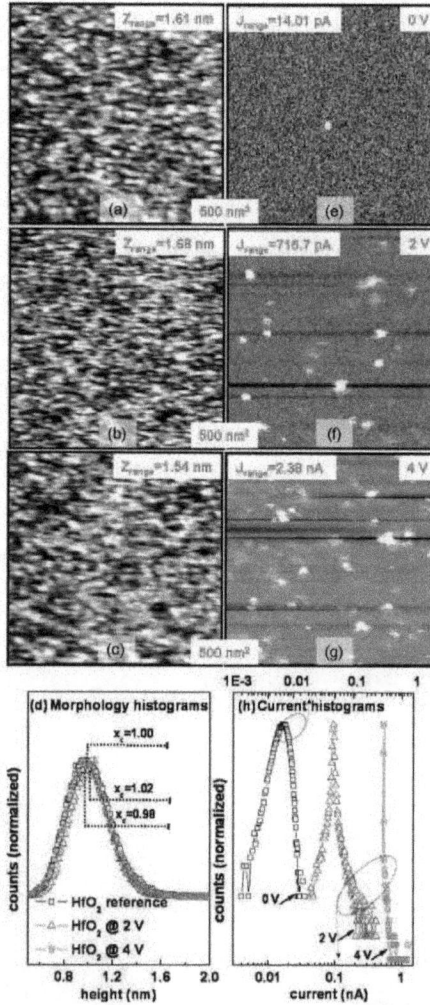

Figure 4.7: C–AFM on 5 nm m–HfO$_2$ film; (a), (b) and (c) 500 nm^2 topology pictures at 0 V, 2 V and 4 V, respectively; d) Morphology histograms at 0 V, 2 V and 4 V; (e), (f), and (g) simultaneously measured leakage current at 0 V, 2 V and 4 V respectively; (h) Current histograms for 0 V, 2 V, 4 V.

60

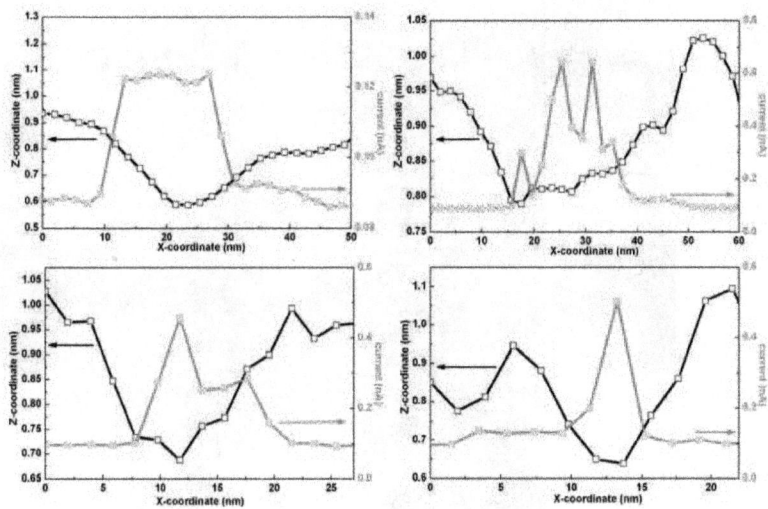

Figure 4.8: C–AFM cross–section scans obtained on 5 nm HfO$_2$ at a voltage of 4 V.

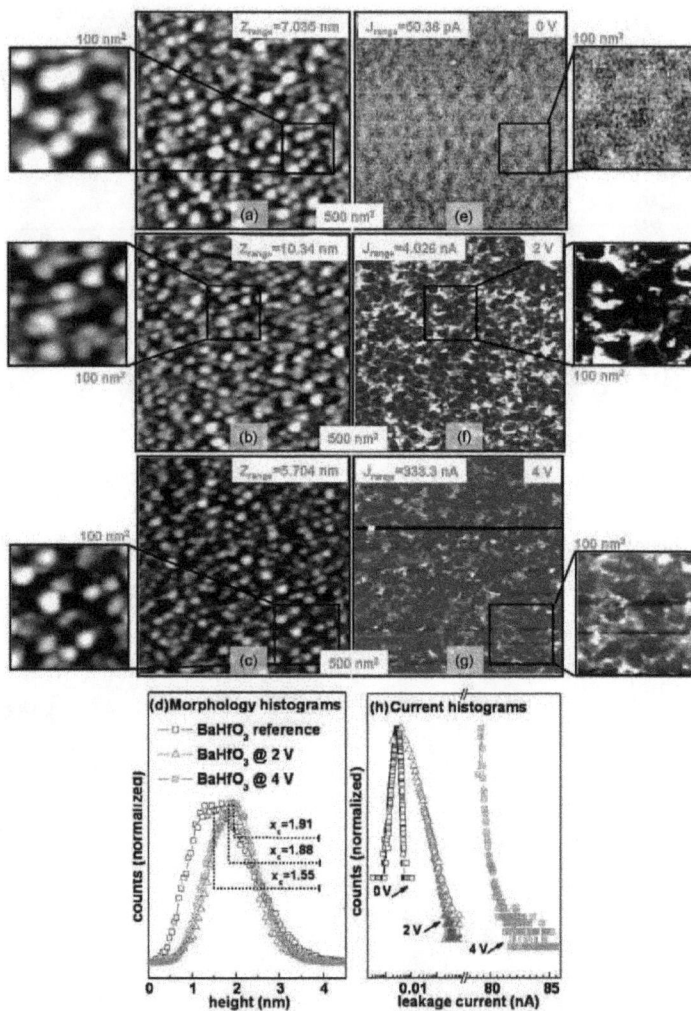

Figure 4.9: C–AFM on 5 nm c–BaHfO₃ film; (a), (b) and (c) 500 nm² topology pictures at 0 V, 2 V and 4 V, respectively; d) Morphology histograms at 0 V, 2 V and 4 V; (e), (f), and (g) simultaneously measured leakage current at 0 V, 2 V and 4 V respectively; (h) Current histograms for 0 V, 2 V, 4 V.

BaHfO$_3$	Z$_{range}$	RMS$_{topo}$	J$_{range}$	RMS$_{leak}$
0 V	7.035 nm	2.277 nm	50.36 pA	20.6 pA
2 V	10.34 nm	2.185 nm	4.026 nA	3.025 nA
4 V	5.704 nm	2.345 nm	333.3 nA	86.2 nA

Table 4: C–AFM acquisition on ~5 nm BaHfO$_3$ layer at 0 V, 2 V and 4 V; topological (Z$_{range}$–maximum topology value, RMS$_{topo}$–root mean square from topology picture) and electrical (J$_{range}$–maximum current, RMS$_{leak}$–root mean square values for current maps) results.

The nanoscopical investigation of HfO$_2$ was followed by a similar analysis based on a ~5 nm thick c–BaHfO$_3$ layer. The results are shown in Fig. 4.9 and summarized in Table 4. The left panel corresponds to the topology at various voltages (0 V, 2 V and 4 V in Fig. 4.9 (a)–(c), respectively) and is followed by respective histograms. The right panel illustrates enlarged data (100 nm^2) from the main measurements in order to discuss the results in detail.

For the reference sample measured at 0 V (Fig. 4.9 (a)), a Z$_{range}$ of 7.035 nm and an RMS of 2.277 nm is obtained. Single grains can be clearly distinguished in this case. One can notice that the grain size for c-BaHfO$_3$ is larger in comparison with the previously presented HfO$_2$ layer (22 nm±2 nm for BaHfO$_3$ vs.16 nm±2 nm for HfO$_2$ obtained from AFM line scans in XY direction). For the measurement acquired at 2 V, (Fig. 4.9 (b)) the values are Z$_{range}$=10.34 nm and RMS=2.185 nm. For the 4 V (Fig. 4.9 (c) measurement, the Z$_{range}$ drops (5.704 nm) whereas the RMS increases (2.345 nm). These values do not show any trend as shown in morphology histograms (Fig. 4.9 (d)), however, it is important to note the change. This may be a result of two processes: surface modification in the reduction–oxidation reaction after voltage application as described by K. Szot [108] or local surface thickness variation. It is noted that the surface modification can be excluded in our case: after each set of measurement the surface was controlled with larger scan area acquisition without bias voltage to see the previously scanned region; this reliability check did not reveal morphological changes with respect to the "fresh" background area. In consequence, we attribute the changes in surface morphology to local thickness variations.

Following the topology description, leakage current maps will be described (Fig. 4.9 (e)–(g)). The 0 V acquisition (Fig. 4.9 (e)) reveals low noise measurement with J_{range}=50pA and RMS_{leak}=20pA. Leakage current appears after 2 V (Fig. 4.9 (f)) application exhibiting a maximal value of J_{range}=4 nA. This is approximately four times larger in comparison with HfO_2 at same measurement conditions. The mean value oscillates here around RMS_{leak}=3 nA. Further voltage increase (4 V in Fig. 4.9 (g)) leads to preamplifier saturation; however this happens only on one certain position (the brightest point on the current map followed by a black shadow along the line). This is probably due to a measurement artifact. Therefore, the average current value RMS_{leak}=86 nA is much lower than the single saturation point. In the histogram below the current maps we see again a Gaussian distribution of the leakage for the reference measurement (0 V). Voltage application (2 V) results in strong contribution from points exhibiting current values around 3 nA. The right skewness of the plot suggests that most points appear above the middle value which suggests local leakage current increase. The whole layer is insulating. However, the grain boundaries formed after Ba introduction, results in a local strong current increase. The average current is one order of magnitude higher than in case of HfO_2. For 4 V measurement, the current changes by two orders of magnitude (~100 nA with respect to HfO_2 values). It is noticed that the same trend of histogram skewness, as for 2 V measurement, is obtained.

Using additional information from the enlarged pictures (left and right $100nm^2$) we can investigate the correlation of leakage current with topology maps (the 0 V acquired noise signal is an exception since there is no information on the current map). For measurements at 2 V and 4 V, a clear current–topology correlation can be observed: the weakest points exist between and around single grains, forming low resistive paths for the current as already suggested before.

The results shown for m-HfO_2 (Fig. 4.7) revealed "hot spots" with higher leakage current. The exact analysis of topology–current correlation led to the conclusion that higher current appeared in thickness reduced regions. For c-$BaHfO_3$ (Fig. 4.8) the leakage current mechanism was different: here the low resistive areas were located at grain boundaries. It is noticed, however, that the grain boundary areas are thinner than the grains. In consequence, it is at present unclear whether a difference in material properties or just a thickness variation is responsible for the higher leakage current.

4.1.2 Conclusions

The study of Ba–added HfO_2 dielectrics on TiN metal electrodes reported in this chapter is summarized in the following:

1. As evidenced by XPS data, simultaneous deposition of BaO and HfO_2 resulted in the formation of homogenous mixed oxide layer with a stoichiometry of approximately $BaHfO_3$

2. According to the XRD data, $BaHfO_3$ deposited at 400 °C is amorphous; further temperature treatment is required for crystallization. RTA at 800 °C triggered $BaHfO_3$ crystallization in the cubic perovskite phase for films >10 nm and confirmed the XPS result with respect to the formation of a homogenous $BaHfO_3$ layer formation.

3. For a–$BaHfO_3$, the dielectric permittivity is 23 and it increased by about 20% with respect to HfO_2 (k~19). Crystallization in c–$BaHfO_3$ by thermal treatment resulted in k~40. As a result of enhanced k values, the CET scaled down to 1 nm. However, the leakage current density increased significantly (from 2.8 10^{-9} A/cm^2 for a–$BaHfO_3$ to $6 \cdot 10^{-7}$ A/cm^2 for c–$BaHfO_3$ with CET=1.4 nm, both measured at 0.5 V). Temperature dependent leakage studies revealed that the leakage current in c–$BaHfO_3$ is governed by the Pool–Frenkel effect.

4. By photoelectron spectroscopy studies, the band offsets for m–HfO_2, a-$BaHfO_3$, and c–$BaHfO_3$ with respect to the TiN metal electrode Fermi level were derived: here, VBO was 3.7 eV, 3.6 eV and 3.7 eV, respectively. For CBO measurements, XAS data were presented: here, 2.0 eV, 2.2 eV and 2.3 eV were obtained, respectively. This results in band gap values of 5.7 eV, 5.8 eV and 6.0 eV for m–HfO_2, a–$BaHfO_3$ and c-BaHfO3, respectively. In other words, the band gap values do not significantly change for these three compounds.

5. C–AFM studies under UHV revealed low resistive "hot spots" in HfO_2 layer: the exact correlation between topology and current revealed higher current in thickness reduced regions. For c-$BaHfO_3$, the leakage mechanism is different: the low resistive areas are located at the grain boundaries.

These results for Ba-added HfO_2 layers prove that $BaHfO_3$ is a promising candidate for future DRAM memory applications in terms of dielectric constant and CET values. However, to fulfill the ITRS requirements for future DRAM generations, further CET scaling and strict leakage control is necessary. In the next chapter of this work, an atomic–scale engineering approach, aiming to improve the electrical parameters of $BaHfO_3$, will be presented.

4.2 Substitution of Hf by Ti ions in BaHfO₃ dielectric layers

Driven by the demand of decreasing CET by increasing the dielectric permittivity and the need for lower crystallization budget, the deposition of stoichiometric $BaHfO_3$ (BHO) was followed by a further material engineering approach. This was accomplished by partial Hf substitution by Ti in $BaHfO_3$. The TiO_2–based compounds have been studied intensively for high–k applications [109, 110] and are attractive because of the expected high permittivity values depending on the crystal structure and deposition method. The high permittivity arises through a strong contribution from soft phonons in Ti-O bonds and is not exhibited by other IVB metal oxides. The substitution of Hf ions for Ti is expected to result in a significant increase of the dielectric constant [111, 112]. For thick films, W. H. Payne et al. reported about a work done on the $BaTiO_3$–$BaHfO_3$ material systems [113]. However, there are only very limited studies available on the properties of very thin $BaHf_{1-x}Ti_xO_3$ films deposited on metallic electrodes.

Figure 4.9 shows the polarization dependence (P) on electrical field (E) applied to compounds with different $BaHf_{1-x}Ti_xO_3$ composition from pure $BaTiO_3$ (x=1) to $BaHf_{0.3}Ti_{0.7}O_3$ (x=0.7).

Compounds with stoichiometry $0.7 \leq x \leq 1$ were reported as ferroelectric dielectrics whereas lower Ti content resulted in a linear P=f (E) relation which is preferred for dielectrics used in DRAM. It is crucial since the paraelectric properties allow using half-V_{dd} method (described in chapter 2.2.1) which reduces the insulator field and avoids reliability problems [49].

Based on the bulk property study, it was our goal to realize paraelectric $BaHf_{1-x}Ti_xO_3$ thin films with x<0.7 and high permittivity. Special focus was devoted to $BaHf_{1-x}Ti_xO_3$ (BHTO) films where half of Hf^{4+} atoms are substituted by Ti^{4+} cations.

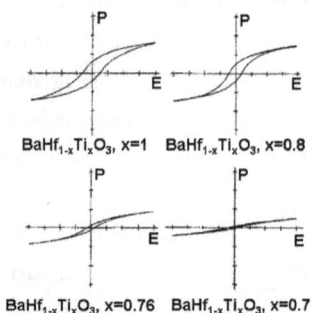

BaHf$_{1-x}$Ti$_x$O$_3$, x=1   BaHf$_{1-x}$Ti$_x$O$_3$, x=0.8

BaHf$_{1-x}$Ti$_x$O$_3$, x=0.76   BaHf$_{1-x}$Ti$_x$O$_3$, x=0.7

Figure 4.9: Hysteresis loops P=f (E) where P–polarization, E–electrical field for different compositions in the BaHf$_{1-x}$Ti$_x$O$_3$ system for 0.7≤x≤1 [114].

4.2.1 Macroscopic study

This part of the thesis is divided into subchapters according to the data which will be discussed: chemical composition, structural properties, electrical characteristics, and band gap investigation.

4.2.1.1 Chemical composition

To achieve the desired stoichiometry, BaO, HfO$_2$ and Ti$_2$O$_3$ fluxes in the MBD chamber were adjusted and controlled using quantitative analysis of XPS spectra. The substitution process is schematically shown in the Fig. 4.10. The composition was confirmed by independent RBS measurements (data not shown).

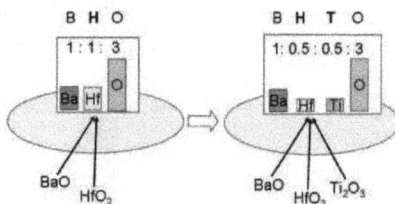

Figure 4.10: Schematical description of the Hf^{4+}–substitution process by Ti^{4+} in BaHfO$_3$ in molecular beam deposition (MBD) chamber.

68

Here, the HfO_2 flux was reduced (50%) to introduce small amounts of the Ti-ions into the BaHfO$_3$ compound and to establish a new dielectric with stoichiometry Ba:(Hf+Ti):O= 1:1:3. The Ti_2O_3 target was chosen because it is easier to evaporate than i.e. TiO or TiO_2. Since the deposition was carried out in oxygen rich environment (oxygen background pressure of 10^{-7} mbar) the targeted Ti^{4+} valence state was achieved.

As a first step after the deposition process, the chemical composition was determined using quantitative in–situ XPS analysis (Fig. 4.11). An example of such a study is summarized in Table 3 (for data evaluation, Eq. 4.6 was used). The data evaluation was carried out according to the procedure discussed in chapter 4.1.1.1.

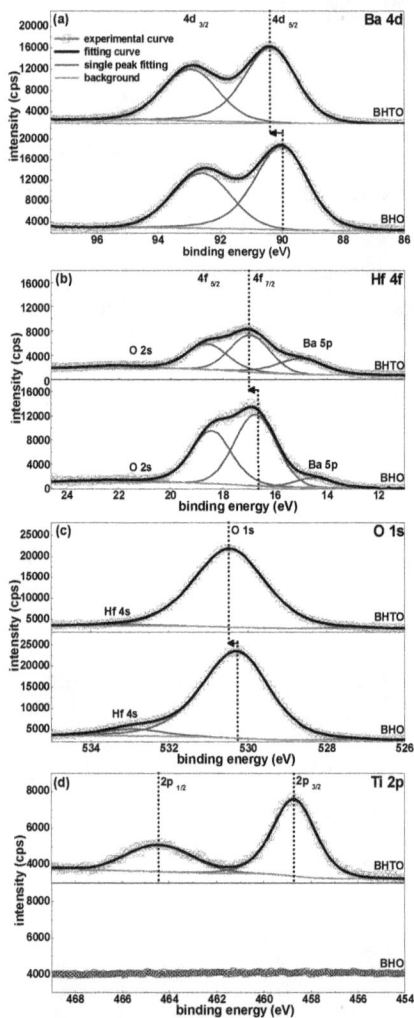

Figure 4.11: XPS study of BHO and BHTO; a) Ba 4d; b) Hf 4f; c) O 1s and d) Ti 2p photoemission lines. The red (print light grey) plot presents the raw data; the black line shows the fitted curve; single peaks are marked in blue (print dark grey) and grey background represents the Shirley function.

Table 5 summarizes the results obtained from XPS spectra (Fig. 4.11). The results are shown separately for each photoelectron line (Ba 4d, Hf 4f, O 1s and Ti 2p) and contain area under photoemission line followed by data normalized by the respective ASF values. Finally, below the single lines, a summary result from the measurement is shown for each compound.

Dielectric compound	Ba 4d (ASF=3.224)	
	XPS peak area	**Area/ASF**
BaHf$_{1-x}$Ti$_x$O$_3$	65042.9	20174.6
	Hf 4f (ASF=2.639)	
	XPS peak area	**Area/ASF**
BaHf$_{1-x}$Ti$_x$O$_3$	26815.41	10161.2
	O 1s (ASF=0.711)	
	XPS peak area	**Area/ASF**
BaHf$_{1-x}$Ti$_x$O$_3$	129322.8	61290.44
	Ti 2p (ASF=2.111)	
	XPS peak area	**Area/ASF**
BaHf$_{1-x}$Ti$_x$O$_3$	6875.37	9670.01
	FINAL RESULT (including ASF)	
BaHf$_{1-x}$Ti$_x$O$_3$	**Ba:Hf:Ti:O = 1 : 0.504:0.479:3.038**	

Table 5: Example of quantitative analysis for BaHf$_{1-x}$Ti$_x$O$_3$ obtained from XPS–area under photoemission line and normalized area with respect to atomic sensitivity factors (ASF); final result indicates a BaHf$_{0.5}$Ti$_{0.5}$O$_3$ stoichiometry.

All samples taken into further consideration exhibited the target BaHf$_{1-x}$Ti$_x$O$_3$ stoichiometry with x≈ 0.5 with a maximal deviation of about 10 % [114].

The analysis of chemical shifts of the Ba 4d, Hf 4f and O 1s and Ti 2p photoemission lines is illustrated in Fig. 4.11.

Figure 4.11 (a) shows the Ba 4d core level emissions of BHO and BHTO. For BHO, two dominant peaks are located at 90.0 eV and 92.5 eV, which are characteristic of the Ba 4d$_{5/2}$ and Ba 4d$_{3/2}$ spin orbit components, respectively, in amorphous

71

BaO-compounds [115]. The spin–orbit splitting (SOS) is 2.5 eV. Ba 4d core level of BHTO shows that addition of Ti ions to the BHO compound results in a shift of the Ba 4d XPS lines towards higher binding energies by $\Delta E = 0.3$ eV (SOS is equal to 2.5 eV as in the BHO).

The Hf 4f photoemission lines are shown in Fig. 4.11 (b). For BHO, two peaks centred at 16.8 eV and 18.5 eV are clearly visible. These signals are characteristic of Hf $4f_{7/2}$ and Hf $4f_{5/2}$ photoemission lines in Hf–compounds, respectively. The binding energy distance of SOS peaks is close to the handbook XPS data [58] and equal to 1.7 eV. For BHTO, the two main peaks are present at 17.1 eV and 18.8 eV for Hf $4f_{7/2}$ and Hf $4f_{5/2}$, respectively. As a result of Ti ion substitution, Hf 4f photoelectron lines shift towards higher binding energy by $\Delta E = 0.3$ eV. In the initial state picture, this can be explained by higher electronegativity of Ti compared to Hf ions (1.54 and 1.3 in Pauling scale, respectively) [92, 95]. A small shoulder on the higher binding energy (~22 eV) can be attributed to O 2s states which overlap with the Hf 4f line. The small peak appearing on the lower binding energy side of Hf $4f_{7/2}$ is due to the Ba 5p photoelectron emission. In comparison with the spectrum for BHO, the intensity of the Hf 4f lines drops after Ti incorporation by about 50 % (extracted from peak area) and the intensity of Ba 5p line remains nearly constant (by otherwise constant experimental conditions).

The corresponding O 1s spectra are shown in Fig. 4.11 (c). For BHO, the main O 1s peak is located at 530.3 eV. At higher binding energy, the Hf 4s photoelectron line appears (533 eV). BHTO reveals the main XPS line at 530.5 eV. Similar to BHO, Hf 4s peak appears on higher binding energy, however, the intensity is much lower here. The main O 1s peak shifts towards higher binding energy in the case of BHTO, resulting in $\Delta E = 0.2$ eV. Replacement of Hf ions with Ti results in an increase of the O 1s binding energy. This is in agreement with literature for compounds containing titanium as for example $BaTiO_3$ (BTO) [116].

Figure 4.11 (d) shows the Ti 2p core level emission. For BHTO, Ti $2p_{3/2}$ and Ti $2p_{1/2}$ peaks are centred at 458.7 eV and 464.4 eV, respectively. Both peak positions and the distance between SOS components, which is close to 5.7 eV, are in agreement with the literature data for fully oxidized Ti^{4+} cation in TiO_2 [117].

The important experimental findings of this part of the study, namely the detection of a 50 % intensity decrease of the Hf lines after substitution by Ti in combination with the detected Ti^{4+} valence state in the dielectric, gives strong evidence for the formation of a homogenous $BaHf_{0.5}Ti_{0.5}O_3$ film in which Ti substitutes the Hf ion in the B–position of the $A^{2+}B^{4+}O_3$ perovskite structure.

4.2.1.2 Structural properties

The deposition and in–situ chemical characterization process was followed by ex–situ investigations. This part summarizes the experimental results obtained by XRD.

Figure 4.12 (a) shows grazing incidence XRD (GIXRD) scans acquired for a ~30 nm thick BHTO sample at various preparation stages. This diffraction technique is very sensitive for thin layers: in our experiment, the x–ray beam hits the surface at grazing incidence (α=0.8°) while the detector angle is scanned over a 2θ range from 24° to 55°.

The as–deposited sample at 400 °C is clearly amorphous as no peaks, apart from the TiN (200) substrate reflection, are visible. Upon RTA treatment, at 600 °C the thin film starts to crystallize exhibiting BHTO (200) and (110) diffraction peaks at 44.6° and 31.1°, respectively. After RTA at 750 °C, the oxide peaks strongly gain intensity which indicates crystallization in the cubic perovskite phase. An additional peak appears on 34.7° which can be ascribed to the monoclinic HfO_2 (020) reflection. This can be an indication for a starting partial decomposition of the dielectric film at high temperatures. The inset in Fig. 4.12 (a) shows a transmission electron microscope (TEM) image of the polycrystalline material after 750 °C RTA treatment, indicating a smooth interface between the dielectric and the TiN electrode.

Figure 4.12(b) shows comparative XRD specular θ–2θ scans of BHO and BHTO films which were subjected to RTA treatments. Temperatures of 800 °C and 600 °C were chosen because the crystallization onset is observed at these temperatures for BHO [32] and BHTO, respectively. The thickness of the dielectric layers is in each case about 30 nm. The results indicate that BHTO crystallizes at lower temperatures compared to BHO. It might be speculated that this is due to the fact that BHTO is an intermediate compound between BHO and BTO. The latter has a rather low crystallization temperature (200 °C-500 °C), where the exact value depends on the deposition method [118,119].

In Figure 4.12(b) both measurements show TiN (200) diffraction peak at 42.7 degree belonging to the underlying electrode. The BHO (200) and BHTO (200) peaks are visible at 43.5 and 44.7 degrees, respectively. Position comparison of the cubic BHO with the cubic BHTO, reveals that the BHTO peak is shifted towards higher 2θ values.

We assume in a first approximation undistorted polycrystalline cubic films and follow Bragg's law in order to calculate the d–spacing between the planes in the atomic lattice and extract thus the cubic unit cell parameter with a= 4.1602 Å and a= 4.0643 Å, for BHO and BHTO, respectively. The difference equals to 2.3 %. Those values are close to powder data for bulk BHO and bulk BHTO (cubic space group Pm3m with a= 4.167 Å and a= 4.0914 Å, respectively).

Based on XRD scans shown in Fig. 4.12(b), the grain size was calculated from the FWHM Bragg peak values [120]. The broadening of the reflections due to strain and instrumentation was neglected in this rough approximation. The grain size was found to be 16 nm for BHO and 6.0 nm for BHTO in the growth direction. This result indicates that 30 nm thick polycrystalline BHO and BHTO films are not composed of columnar grains extending over the whole film thickness but rather of various grains stacked in the growth direction.

Figure 4.12 (c) summarizes the lattice parameter discussion for BHO and BHTO. Powder diffraction parameters for bulk films and experimental data for thin layers as a function of titanium concentration are plotted. Experimental data obtained for bulk and thin cubic BTO films are included for completeness [121]. A linear relation exists between the crystal lattice parameters of those alloys and the Ti metal concentration. According to Vegard's law [122], the experimentally observed change in the lattice constant parameter between BHO and BHTO indicates that about 50 % of Hf^{4+} was substituted by Ti^{4+}, a result fully in line with the above quantitative XPS study. Difference between values for bulk material and thin films (averaged for BHO, BHTO and BTO ~0.56 %) is visible on the plot. A possible origin might be given by strain in the dielectric material.

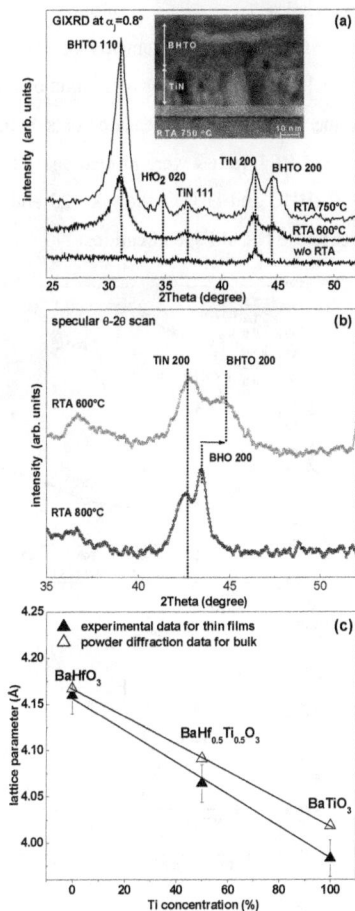

Figure 4.12: XRD analysis from BHO and BHTO; a) GIXRD for BHTO as–deposited and RTA treated at 600° C and 750° C; the inset shows an XTEM image of BHTO layer after RTA at 750°C; b) Specular θ–2θ scans for BHO at 800 °C and BHTO at 600 °C; c) Lattice parameter extracted from θ–2θ scans as a function of Ti–concentration, experimental data for BaTiO₃ were obtained from powder diffraction data base, the error bars show 5 % deviation.

4.2.1.3 Electrical characteristics

We now present the results of electrical measurements for Pt/high–k/TiN MIM structures. Figure 4.13(a) shows the dielectric constant k as a function of RTA temperature for BHO and BHTO. To extract the k values, CV measurements in the range of -1÷1V and 100 kHz were performed and CET values were plotted against physical thickness [123]. After deposition at 400 °C, BHO and BHTO show k ~23 and ~18, respectively. These values are too low to meet future DRAM specifications.

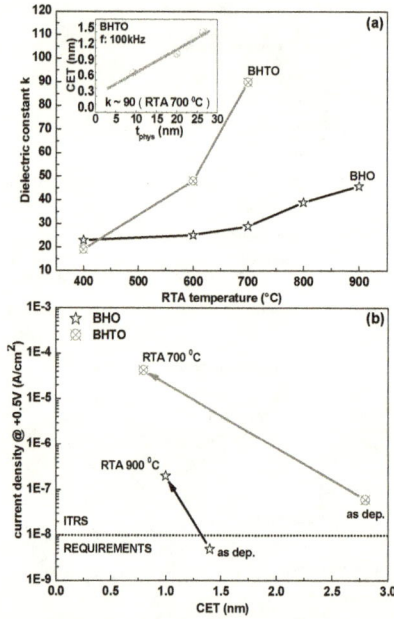

Figure 4.13: Electrical characteristics of Pt/high–k/TiN MIM structures; a) Dielectric constant dependence on RTA treatment for BHO and BHTO, (inset-capacitance equivalent thickness (CET) as a function of physical thickness for BHTO); b) Leakage current density at 0.5 V as a function of CET for BHO and BHTO. As deposited amorphous samples are compared with polycrystalline layers for BHTO and BHO after RTA at 700 °C and 900 °C, respectively. Dashed line indicates ITRS requirements.

Subsequent RTA increases k so that at 700 °C values of 30 for BHO and 90 for BHTO are obtained. In consequence, substitution of Hf^{4+} ions by Ti^{4+} results in BHTO to a three times higher dielectric constant as compared to BHO.

In the inset, a plot of capacitance equivalent thickness (CET) as a function of physical thickness for three BHTO samples is shown. As discussed above, from the slope of this plot we can extract the dielectric constant value of about 90 for BHTO. The intercept of around 0.3 nm CET indicates the existence of an interfacial low–k layer between the high–k BHTO layer and the TiN metal electrode. A future non–destructive SR–XPS study (similar to a work on $Pr_xAl_{2-x}O_3$ (x=0–2)/TiN systems) is recommended for further investigation of the nature of this layer [124].

Figure 4.13(b) illustrates the leakage current density extracted from J–V characteristics at 0.5 V as a function of CET. As–deposited and annealed samples of BHO and BHTO at 900 and 700 °C, respectively, are compared. For as-deposited BHO we obtained current density of $5.1 \cdot 10^{-9}$ A/cm^2 at 1.4 nm CET (dashed line marks the ITRS requirement $J_{leak} < 10^{-8}$ A/cm^2 at 0.5 V for CET 0.3-0.5 nm). After annealing at 900 °C, the CET scales down to ~1 nm and the current density increases to $2 \cdot 10^{-7}$ A/cm^2. For amorphous BHTO, we detect a leakage current value of $6.1 \cdot 10^{-8}$ A/cm^2 at 0.5 V and 2.8 nm CET. This quite high value can be due to low conduction band offset (CBO) of BHTO on TiN which will be discussed in the next subchapter. Crystallization of the film upon RTA treatment results in a decrease of CET to ~0.8 nm but an abrupt increase in the leakage current is observed ($4 \cdot 10^{-5}$ A/cm^2 at 0.5 V). The increase in leakage current can be influenced by two factors: the already mentioned low CBO in combination with grain boundary governed leakage. An approach to study the latter one will be presented in the nanoscopical C-AFM study chapter discussed further below.

The electrical characterization suggests an advantage of using Ti–doped BaHfO$_3$ layers with respect to BHO. The film shows promising dielectric permittivity and, what is important for processing, lower thermal budgets. On the other hand, the leakage current density regime is too high with respect to the target values. This indicates that an approach for optimizing the leakage properties of thin BaHf$_{0.5}$Ti$_{0.5}$O$_3$ films on TiN metal electrodes is necessary.

In summary, due to Hf^{4+} substitution with Ti^{4+}, the dielectric permittivity increased after RTA from 30 to 90 and the CET value was accordingly lowered. However, at the same time the leakage current density increased substantially by the required RTA treatment. An increased leakage current density can be caused by an improper band alignment at the dielectric–metal interface and grain boundary formation after temperature treatments [125, 126]. To address the band gap alignment aspect, we performed a study on the electronic properties (i.e. band gap values and band alignment) presented in the next sections.

4.2.1.4 Band gap and band alignment

To gain insight into the band alignment in BHTO with respect to TiN, we performed synchrotron–based XPS (hv=200 eV) combined with respective spectroscopic ellipsometry (SE) study and synchrotron-based XAS (SR-XAS) measurements. For comparison, the band alignment in BHO was included.

The XPS spectra plotted in the Fig. 4.14 (a) show valence band (VB) spectra for c-BHO, a-BHTO, and c-BHTO. For c–BHO, the VB maximum (VBM) formed mainly by O 2p states is located at around E_B=3.7 eV below the TiN Fermi level (E_F at E_B=0 V). Substitution of Hf^{4+} ions in the cubic perovskite by Ti^{4+} does not introduce dramatic changes neither in the shape of the VB spectra, nor in the position of the VBM which for a–BHTO and c–BHTO is located at E_B=3.4 eV and E_B=3.3 eV, respectively. This is in a good agreement with theoretical and experimental explanations, as VBM is formed in the oxides mainly by O 2p states [65].

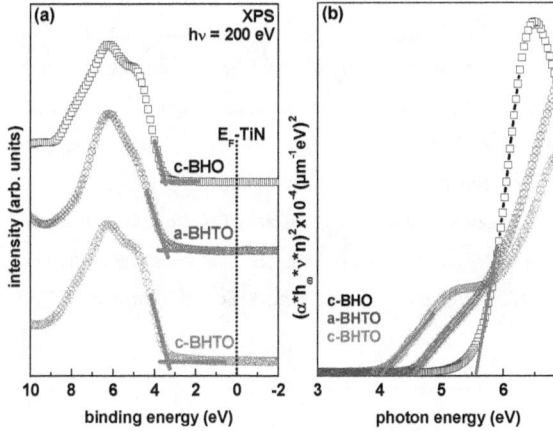

Figure 4.14: Band structure of c–BHO, a–BHTO and c–BHTO, a) XPS spectra; b) Ellipsometry measurements.

Band gaps of BHO and BHTO were measured using the SE's parameters discussed below. The optical constants, refractive index (n), and extinction coefficient (k) are determined by a combination of SE at 70° angle of incidence and normal incidence reflectometry. The dependence of the absorption coefficient α ($\alpha=4\pi k/\lambda$) on photon energy $h\nu$ has been evaluated according to the relation:

$$\alpha \cdot h\upsilon \cdot n = A(h\upsilon - E_g)^{1/2} \tag{4.12}$$

where E_g is the optical band gap energy [127]. The constant A contains the oscillator strength or the optical matrix element and the reduced effective mass. The formula was derived by E. J. Johnnson [128] and R. A. Smith [129], assuming direct interband transitions between the valence and conduction band with simple parabolic bands. This is a reasonable approach to determine the band gaps of BHO and BHTO in our case because we expect the VBM–conduction band minimum (CBM) transition from O 2p to metal d states (O 2p-Hf 5d for BHO and O 2p-Ti 3d for BHTO) to be dipole allowed ($\Delta l = \pm 1$). [130].

79

The results of E_g measurements are shown in Fig. 4.14 (b). The optical band gap of c–BHO is about ~5.8 eV, of a–BHTO about ~4.6 eV, and that of c–BHTO ~4.0 eV.

Table 4 shows the summary of XPS and SE measurements. Additionally, a graphical representation of the band offset and band gap alignment was included below the table. We combined the XPS and SE data to estimate the band alignment of BHO and BHTO on TiN. It is seen that after the substitution of Hf ions by Ti, the VB offset remains almost unchanged whereas the CB offset (CBO) shrinks drastically from 2.1 eV to the value of about 0.7 eV. This decrease in CBM values is due to the fact that, with respect to BHO, not Hf 5d but typically lower lying Ti 3d states form the CBM [46,99].

	VBO	BG (Ellipsometry)	CBO=BG–VBO
c–BaHfO₃	3.7	5.8	2.1
a–BaHf₀.₅Ti₀.₅O₃	3.4	4.6	1.2
c–BaHf₀.₅Ti₀.₅O₃	3.3	4.0	0.7

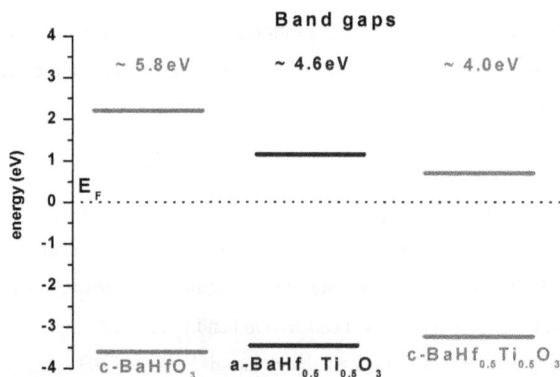

Table 4: Band offsets and band gap determination based on XPS and SE; notation: VBO-valence band offset, CBO–conduction band offset, BG–band gap; below the table, a graphical representation of band gaps and band offsets is shown.

Similar trends were experimentally observed for $BaHf_{1-x}Ti_xO_3$ layers using XPS and X-ray absorption spectroscopy (XAS) [131].

Figure 4.15 compares a combined XPS (hv=200 eV) in (a) and XAS (b) measurement for c-BHO and c-BHTO. The VBM are formed in both cases by O 2p states. For c-BHO, the investigated offset is located around 3.7 eV below the TiN Fermi level. The corresponding value for c-BHTO is 3.3 eV. It is therefore concluded that the VBO value for c-BHTO does not change significantly with respect to c-BHO.

To study the CBO of the materials under investigation, the lowest unoccupied levels in the CBO region were monitored by XAS O K-edge measurements (Fig. 4.15 (b)). The electron is excited from the O 1s core state into empty O 2p states (dipole-allowed transition) and hybridized with metal d states mainly forming the conduction band. The location of the CBO can then be derived by the energy difference between the O 1s XAS peak and the O 1s XPS peak. [106]. In our case, O 2p states hybridize with Hf 5d states in BHO and O 2p with Ti 3d states in BHTO as marked in Fig. 4.15 (b).

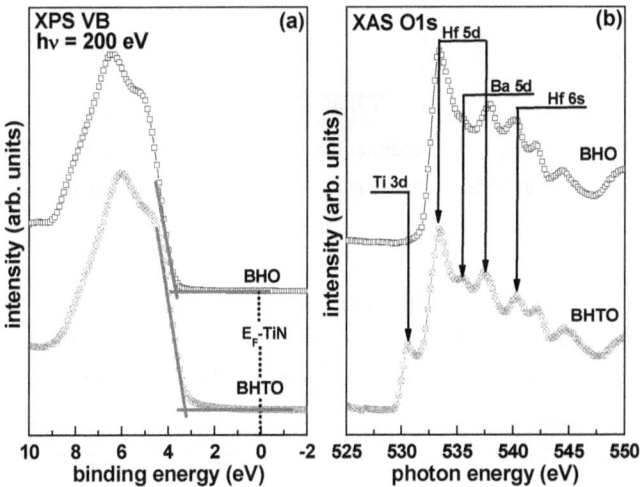

Figure 4.15: Band structure of c–BHO and c–BHTO, a) XPS spectra; b) XAS measurement.

We compared the XAS scan for c-BHO and c-BHTO. In the case of c-BHO, the CBO is formed mainly by Hf 5d metal states and therefore the value of 2.2 eV is obtained. The situation changes dramatically after Ti ions incorporation-the CBO shifts to the value of 0.6 eV. This is caused by lower lying Ti 3d metal states as shown in the spectra.

	VBO	XPS	XAS	CBO	BG
c–BaHfO$_3$	3.7	530.3	532.5	2.2	5.9
c–BaHf$_{0.5}$Ti$_{0.5}$O$_3$	3.3	530.5	531.1	0.6	3.9

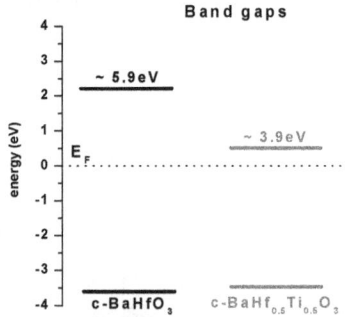

Table 5: Band offsets and band gap determination based on XPS and XAS; notation: VBO–valence band offset, CBO–conduction band offset, BG–band gap; below the table, a graphical representation of band gaps and band offsets is shown.

Table 5 summarizes the results obtained from combined XPS-XAS measurements. It is followed by a graphical band offset representation. Here we obtain the values of 5.9 eV and 3.9 eV for c-BHO and c-BHTO, respectively. The most pronouncing difference is visible in the CBO region.

Such low CBO in c-BHTO is likely to be the source of leakage problems. As a result, high work function electrodes might be required in future to improve the CBO and in consequence the leakage characteristics of BHTO thin films.

4.2.2 Nanoscopical investigation

In this part of the thesis, we focus on the leakage current behaviour of BHTO samples investigated on the nanoscale using C–AFM. Measurements for a 10 nm thin sample, subjected to different thermal treatments, are compared. We first discuss the topology results (Fig. 4.16 (a) and (b)) and the current maps (Fig. 4.16 (c) and (d)). At the end, we focus on the correlation between topology and current images.

Figure 4.16 (a) shows topology images obtained on $1 \mu m^2$ scan area for the as–deposited sample and after RTA treatments at 600 °C and 700 °C.

In the morphology maps, we see white and dark areas corresponding to high and low feature in the topography, respectively. With increasing annealing temperature, the numbers of large hillocks and in consequence the inhomogeneity of the film decreases since the maps become more uniform.

Figure 4.16 (b) presents histograms extracted from the topology height. The histogram distribution is similar for all samples. This distribution reflects the root mean square (RMS) values of the roughness which are in consequence almost identical for the three images (as-deposited 1.71 nm, 600 °C: 1.74 nm and 700 °C: 1.66 nm). More interestingly, the middle point of the curve, x_c, shifts toward smaller values with increasing RTA temperature. The difference between these average values x_c is >2 nm. With an absolute film thickness of ~10 nm, this change in the film morphology can certainly strongly contribute to different leakage current behaviour, as discussed in the following.

Figure 4.16 (c) shows nanoscopical current maps acquired parallel to the topology. Bright spots represent high leakage current and dark points refer to low current values. With increasing RTA temperature, the number of low resistive points increases and the current map becomes less uniform.

Figure 4.16 (d) presents current histograms in the lin–log (as deposited film) and log–log (RTA treated samples) scale. The number of scanned points with constant voltage (4 V) is plotted as a function of the detected leakage current.

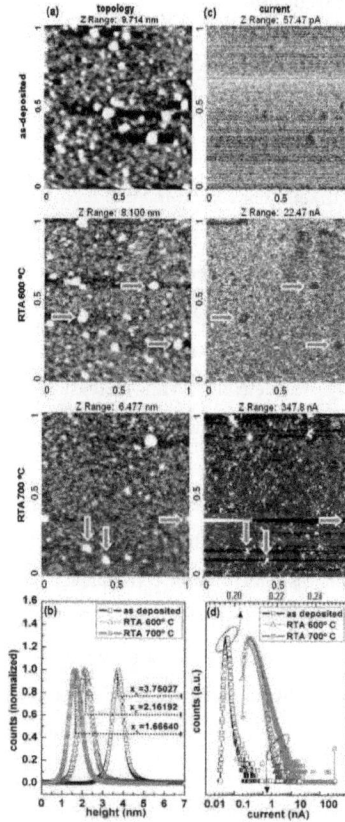

Figure 4.16: C–AFM measurements on BHTO as deposited and after RTA treatments at 600 and 700° C; a) topology images, 1μm² size. The white points represent high morphology structures. The Z–range refers to the maximum values from each measurement. Arrows and circles refer to the correlation with current maps; b) Height histograms for three thermal treatments (data normalized to 1); c) current images at 4 V bias acquired parallel to the topology, white positions represent highest current, d) Current histograms in the lin–log scale for as deposited sample (upper axis) compared with two RTA temperatures in the log–log scale (bottom axis). Data extracted from DC current maps.

84

For the as–deposited sample, we obtain a very low current signal which is close to the noise level of the C–AFM instrument. For the sample annealed at 600 °C, we obtained uniformly distributed current in the nA–range. For sample annealed at 700 °C, a broader distribution of current is observed. Most points are in the range <50 nA. Single features appear at lower and higher current values; however the amount of high leakage points, so called "hot spots", is larger. This is indicated on the plot by the number of points at higher current values.

Comparison of topology and current data suggest a correlation between morphology and leakage current distribution (the as–deposited sample is an exception where a well known piezo "creep" or "wrap" effect appears [132]).

For the sample treated at 600 °C, we observe that higher features on the topology image (some examples are marked with arrows) correspond to low current positions. However, this morphology–current correlation showing that thick film area correspond to low leakage regions is not always that straightforward. This was marked with red circles showing no correspondence of topology and current maps.

For the film annealed at 700 °C we can distinguish three "hot spots" on the 1 μm^2 scale (marked with arrows) which mostly contribute to the high current range. As seen in the topology maps, these "hot spots" in the current correspond to hillocks areas with an average size of about 50 nm. We suspect local nonstoichiometries within the dielectric thin film to be the origin of this high leakage current behaviour, as previously reported for C–AFM studies on $BaZrO_3$ [133].

In summary, the AFM topology becomes smoother as the film is subjected to thermal treatment. As a result of film densification, the leakage current increases. Additionally, the current magnitude is strongly influenced by the formation of "hot spots". In general, the nanoscopic C–AFM studies confirm the macroscopic J–V measurements trend, namely the leakage current increase as a function of RTA temperature for BHTO system on TiN electrode.

4.2.3 Conclusions

The study of Ti–added BaHfO$_3$ layers on TiN metal electrode reported in this subchapter is summarized in the following.

1. As evidenced by the quantitative XPS study, the substitution of Hf^{4+} by Ti^{4+} ions in BaHfO$_3$ resulted in the formation of BaHf$_{1-x}$Ti$_x$O$_3$ layer. The detection of 50 % intensity decrease of the Hf 4f photoemission line after substitution by Ti in combination with the detected Ti^{4+} valence state in the dielectric gives strong evidence for formation a homogenous BaHf$_{0.5}$Ti$_{0.5}$O$_3$ film stoichiometry.

2. The complementary XRD study revealed lowered crystallization temperature for BaHf$_{0.5}$Ti$_{0.5}$O$_3$ (600 °C) compound which renders this dielectric compatible with the thermal budget of DRAM processes. The experimentally observed change in the lattice constant parameter between BaHfO$_3$ and BaHf$_{0.5}$Ti$_{0.5}$O$_3$ corroborates the fact that about 50 % of Hf^{4+} was substituted by Ti^{4+} (which is in line with the quantitative XPS study).

3. The electrical data showed that, due to Hf^{4+} substitution with Ti^{4+}, the dielectric permittivity increased after RTA treatment at 700 °C three times (k~90) with respect to BaHfO$_3$ (k~30) at the same conditions whereas the CET values were accordingly scaled down to 0.8 nm. An increased leakage current density was observed for BaHf$_{0.5}$Ti$_{0.5}$O$_3$ which was attributed to film inhomogeneities (grain boundaries etc.) influence or a low CBO at the dielectric–metal interface.

4. The combination of XPS and ellipsometry measurements revealed that substitution of Hf by Ti ions did not significantly influence VBO whereas the CBO shrunk drastically from 2.1 eV to the value of 0.7 eV. This is because lower lying Ti 3d states with respect to Hf 5d states form the CBM. Similar values were obtained by a combination of XPS and XAS techniques. Such low values are likely to be the source of leakage problems. As a future step, high work function electrodes are required which could improve the electrical leakage characteristics.

5. The macroscopic study was corroborated by nanoscopical study based on C–AFM measurements. The topology became smoother, as the film was subjected to thermal treatment. As a result of the crystallization process, the leakage current increased which is the effect of "hot spots" formation associated with local film

86

non–stoichiometries and/or by the formation of grain boundaries introducing low resistive leakage paths.

In conclusion, the Ti–doped layers show a promising approach for improving the dielectric properties of $BaHfO_3$ in view of future DRAM storage applications. Further investigation should focus on electrodes with higher work functions (i.e. Ru or RuO_2). Furthermore, the deposition under fully professional Si–cleanroom condition probably leads to more homogenous film thicknesses, improving further the leakage characteristics.

Chapter 5

Summary and outlook

5.1 Summary of technical achievements

The goal of this research study was to evaluate the potential of Si CMOS compatible HfO_2-based dielectric layers for future DRAM applications. For this purpose, the materials science and dielectric properties of HfO_2 need to be further developed. In this work, we prepared the following two materials science systems:

1) Ba-added $HfO_2 \rightarrow BaHfO_3$

A simultaneous deposition of BaO and HfO_2 led to the $BaHfO_3$ dielectric. The electrical data revealed increased dielectric constant with respect to HfO_2. The band gap investigation showed that addition of BaO to HfO_2 does not influence the band gap values in $BaHfO_3$ and band offset with respect to TiN metal electrode.

2) Ti-added $BaHfO_3 \rightarrow BaHf_{0.5}Ti_{0.5}O_3$

Hf^{4+} ions substitution by Ti^{4+} in $BaHfO_3$ resulted in an increased dielectric constant and lowered thermal budget. Meanwhile, the leakage current density increased what was attributed to improper band alignment and film thickness inhomogeneities.

In the following, the main results are summarized in more detail.

$BaHfO_3$ vs. HfO_2 dielectric films

Figure 5.1 shows the monoclinic structure of HfO_2 (a) and cubic cell of $BaHfO_3$ (b) obtained in this study. The HfO_2 cell consists of hafnium atoms (blue) and surrounding oxygen atoms (red). For $BaHfO_3$, an additional barium central atom (grey) is visible.

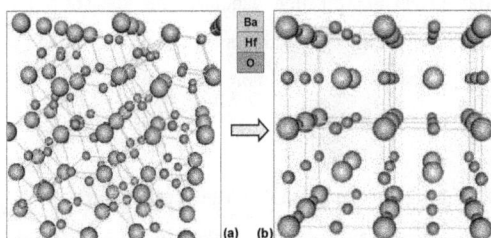

Figure 5.1: Unit cell structure of m-HfO$_2$ (a) and c-BaHfO$_3$ (b).

Figure 5.2 compares the electrical (a) and electronic data (b-c) for m-HfO$_2$, a-BaHfO$_3$ and c-BaHfO$_3$. The electrical data indicate an increase of dielectric constant from 19 for HfO$_2$ to 23 for a-BaHfO$_3$. Crystallization of the BaHfO$_3$ compound leads to further increase of the dielectric constant to about 38. The CET values are scaled down to 1 nm with respect to HfO$_2$ (~1.5 nm). Meanwhile, the electronic structure (Fig. 5.2 (b)) shows that the VBO does not change with respect to TiN metal electrode. Barium addition to HfO$_2$ does also not influence the CBO formed by Hf 5d metal states.

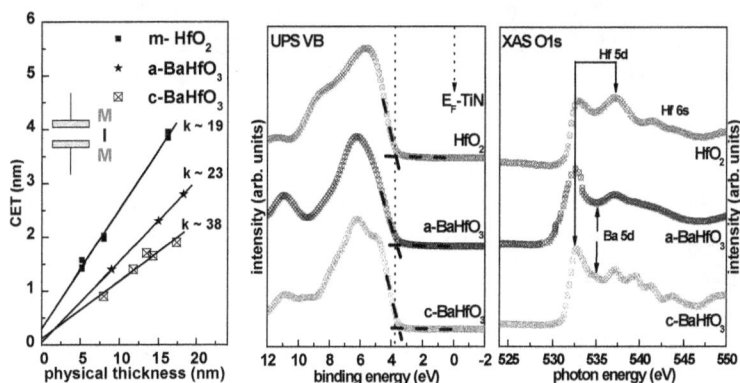

Figure 5.2: Comparison of electrical data for m-HfO$_2$, a-BaHfO$_3$ and c-BaHfO$_3$. The electrical data represented by CET=f(physical thickness) in (a) and electronic structure with respect to valence band offset (b) and conduction band offset (c).

89

In conclusion, BaHfO₃ is a promising candidate i.e. for DRAM trench technology applications in terms of increased dielectric constant and decreased CET values. A very nice result is given by the fact that the improvement of dielectric permittivity does not influence the band gap of this material.

Substitution of Hf by Ti ions in BaHfO₃ dielectric layers

Especially for DRAM stack technologies, thin dielectric films with higher k-values are required. A further engineering approach of BaHfO₃ layers was undertaken by substitution of Hf^{4+} ions by Ti^{4+}, schematically shown in Fig. 5.3. Additional Ti atoms (green) appear on the positions of Hf atoms since the goal was to substitute half of Hf^{4+} ions by Ti^{4+} on the B position of the $A^{II+}B^{IV+}O_3$ perovskite structure.

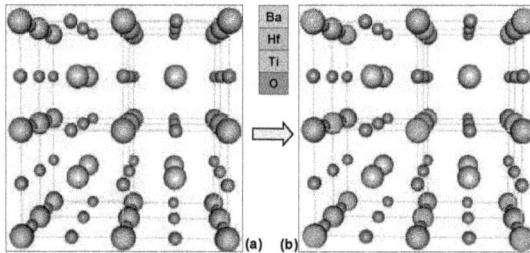

Figure 5.3: Unit cell structure of c-BaHfO₃ (a) and c-BaHf$_{0.5}$Ti$_{0.5}$O₃ (b).

Having formed a stoichiometric BaHf$_{0.5}$Ti$_{0.5}$O₃ compound (proven by XPS and XRD) we investigated the electrical and electronic properties of the newly formed compound and compared it with respect to BaHfO₃.

Figure 5.4 shows a comparison of electrical (a) and electronic (b-c) parameters of c-BaHfO₃ and c-BaHf$_{0.5}$Ti$_{0.5}$O₃. The dielectric constant was further increased from 38 for c-BaHfO₃ to about 90 for c-BaHf$_{0.5}$Ti$_{0.5}$O₃. The thermal budget was lowered from 900 °C to 700 °C which renders the dielectric more compatible with Si technology processes. The CET values were scaled down to about 0.8 nm, respectively.

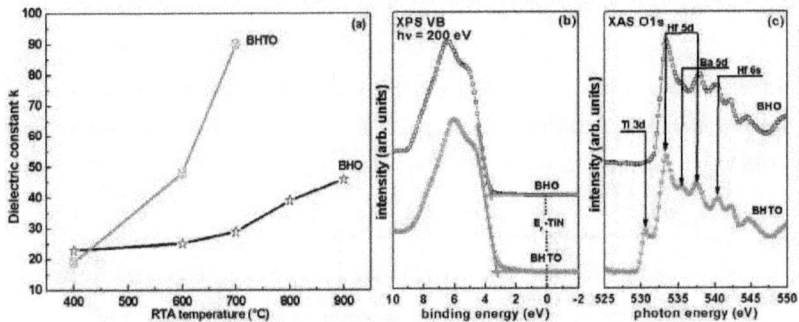

Figure 5.4: Comparison of electrical data for c-BaHfO$_3$ and c-BaHf$_{0.5}$Ti$_{0.5}$ O$_3$. The electrical data are represented by CET versus physical thickness in (a), electronic structure with respect to valence band offset in (b) and conduction band offset in (c).

To gain insight into the electronic structure of the material and band alignment issue, a combined XPS/XAS investigation was performed. Fig. 5.4 (b) shows only slight change of VBO after Ti incorporation into BaHfO$_3$ with respect to the Fermi level of the TiN metal electrode. A significant change was observed in the CBO region where the values shifted from 2.1 eV for BaHfO$_3$ to 0.7 eV in the case of BaHf$_{0.5}$Ti$_{0.5}$O$_3$. In addition, the C-AFM measurements revealed inhomogeneities ("hot spots") in the current maps. Therefore, after Ti incorporation into BaHfO$_3$, an abrupt increase of leakage current was observed. A combination of high work function electrodes with homogenous thin film dielectric deposition process could solve this problem in future.

5.2   Outlook and future activities

Table 6 summarizes the results obtained in this research study starting with the HfO_2 as a mainstream material which is compared to $BaHfO_3$, $BaHf_{0.5}Ti_{0.5}O_3$ and finally to the ITRS requirements for the year 2014.

	RTA	k value	CET	J_{leak} at 0.5V
HfO_2	no	19	>1.5 nm	$3x10^{-9} A/cm^2$
$BaHfO_3$	900 °C	38	1.0 nm	$6x10^{-7} A/cm^2$
$BaHf_{0.5}Ti_{0.5}O_3$	700 °C	90	0.8 nm	$4x10^{-5} A/cm^2$
ITRS for 2014	600 °C	100	0.4 nm	$1x10^{-8} A/cm^2$

Table 6: Parameters outlook compared to ITRS requirements for the year 2014, green color (print light grey) -requirements fulfilled, red color (print black) -requirements not fulfilled, yellow color (print dark grey) -requirements close to be fulfilled.

$BaHfO_3$:
Ba addition to HfO_2 leads to an increased dielectric constant and lowered CET values. The band structure investigation revealed no influence on the band alignment after Ba incorporation.

$BaHf_{0.5}Ti_{0.5}O_3$:
Atomic-scale engineering of $BaHfO_3$ by Ti addition leads to a significant improvement of the dielectric parameters. However, there are issues to be further optimized: the CET value must be further scaled down and the leakage current substantially decreased.

Due to our industrial partner insolvency (Qimonda), the activities with respect to DRAM application will however not be continued at IHP.

Further investigation in our group on these materials will consider different MIM applications in the Back End of Line (BEOL) of IHP's BiCMOS technology, i.e. passive MIMs for radio frequency (RF) applications, embedded NVM as resistive random access memory (RRAM) cells or even RF resonant tunneling diodes (RTD). The technical specifications for the above analog devices are however different than for digital elements. The films can be grown physically thicker and thus the leakage current requirements are not as strict as in DRAM applications.

Literature

[1] *Voijn G. Oklobdzija*, "Digital Design and Fabrication", The computer Engineering Handbook, Second Edition, 5.1-5.6, (1997).

[2] http://www.intel.com/technology/silicon/micron.htm

[3] *B. El-Kareh, G. B. Bronner*, and *S. E. Schuster*, Solid State Technol., 40(5), 89, (May 1997)

[4] E.W. Pugh, D. L. Critchlow, R. A. Henle, and L. A. Russell, IBM Journal of Research and Development, **25**, 5, 585-602, (September 1981)

[5] http://www.intel.com/pressroom/kits/events/moores_law_40th/

[6] THE INTERNATIONAL TECHNOLOGY ROADMAP FOR SEMICONDUCTORS: (2007), www.itrs.net

[7] T. Mikolajick, Infineon Technologies, 6th Dresdner Sommerschule Mikroelektronik, own CD materials, (2007)

[8] *Jerry C. Whitaker,* "The electronics handbook", 1445-1454, (1995)

[9] W. Mueller *et al.*, IEDM Tech. Digest, 339 (2005)

[10] "NEC Electronics Unveils 90nm Embedded DRAM Technology," see http://www.necel.com/news/en/archive/0503/0701.html

[11] W. *Rösner, T. Schulz, L. Risch, F. Hofmann,* "DRAM cell circuit", United States Patent 6362502, (2002)

[12] J.A. Mandelman *et al.,* IBM J. Res. & Dev., **46**, 2/3, (march/may 2002)

[13] W.Noble and W. Walker, "Fundamental, IEEE Circuits & Devices Magazine **1**, 45-51 (1985)

[14] H. Sunami, T. Kure, N.Hashimoto, K. Itoh, T. Toyabe and S.Asai, IEDM, 806-808 (1982)

[15] B. Davari, C. W. Koburger, R. Schulz, J. D. Warnock, T. Furukawa, M. Jost, Y. Taur, W. G. Schwittek, J. K.DeBrosse, M. L. Kerbaugh, and J. L. Mauer, IEDM Tech. Digest, 861 (1989)

[16] D. Kenney, P. Parries, P. Pan, W. Tonti, W. Cote, S.Dash, P. Lorenz, W. Arden, R. Mohler, S. Roehl, A.Bryant, W. Haensch, B. Hoffman, M. Levy, A. J. Yu, and C. Zeller, IEEE Symposium on VLSI Technology, Digest of Tech. Papers, 14-15 (1992)

[17] L. Nesbit, J. Alsmeier, B. Chen, J. DeBrosse, P. Fahey, M. Gall, J. Gambino, S. Gernhardt, H. Ishiuchi, R.Kleinhenz, J. Mandelman, T. Mii, M. Morikado, A. Nitayama, S. Parke, H. Wong, and G. Bronner, IEDM Tech. Digest, 627– 630, (1993)

[18] L. Nesbit, J. Alsmeier, B. Chen, J. DeBrosse, P. Fahey, M. Gall, J. Gambino, S. Gernhardt, H. Ishiuch, R. Kleinhenz, J. Mandelman, T. Mii, M. Morikado, A. Nitayama, S. Parke, H. Wong, and G. Bronner, IEDM Tech.Digest, 627, (1993)

[19] G.B. Bronner, IEDM Symposium, A, Hsinchu, Taiwan, 75-82, (December 1996)

[20] H.S. Kim, D.H. Kim Park, J.M. Hwang, Y.S. Huh, M. Hwang, H.K. Kang, N.J. Lee, B.H. Cho, M.H. Kim, S.E. Kim, J.Y. Park, B.J. Lee, J.W. Kim, D.I. Jeong, M.Y. Kim, H.J. Park, Y.J. Kinam Kim, IEEE, 17.2.1- 17.2.4 (2003)

[21] T. Tran, R. Weis, A. Sieck, T. Hecht, G. Aichmayr, M. Goldbach, P.-F. Wang, A. Thies, G. Wedler, J. Nuetzel, D. Wu, C. Eckl, R. Duschl, T.-M. Kuo, Y.-T. Chiang, W. Mueller, Electron Devices Meeting, IEDM, 1-4, (2006)

[22] W. Müller, "DRAM Scaling Roadmap to 40nm", IEDM (2005)

[23] T. Schroeder, G. Lupina, R. Sohal, G. Lippert, Ch. Wenger, O. Seifarth, M. Tallarida, and D. Schmeisser, J. Appl. Phys. **102**, 014103 (2007)

[24] J. A. Mandelman, R. H. Dennard, G. B. Bronner, J. K. DeBrosse, R. Divakaruni, Y. Li, C. J. Radens, IBM J. Res. and Dev., **46**, No. 2/3, (Mar/May 2002)

[25] C. W. Teng. DRAM Technology Trend. Proc. of the 1995 International Symposium on VLSI Technology, Systems and Applications, 295-299, (1995)

[26] J.G. Simmons, J. Appl. Phys., **34**, 1793, (1963)

[27] T. Schroeder G. Lupina, R. Sohal, G. Lippert, Ch. Wenger, O. Seifarth, M. Tallarida, and D. Schmeisser, J. Appl. Phys. 102, 014103, (2007)

[28] J. Dabrowski, S. Miyazaki, S. Inumiya, G. Kozlowski, G. Lippert, G. Lupina, Y. Nara, H-J. Müssig, A. Ohta, and Y. Pei, „The influence of defects and impurities on electrical properties of high-k dielectrics", in Advances in Electronic Materials, Trans Tech Publications (2009)

[29] E. Gerritsen, N. Emonet, Ch. Caillat, N. Jourdan, M. Piazza, D. Fraboulet, B. Boeck, A. Berthelot, S. Smith, P. Mazoyer, Solid-State Electronics **49**, 1767-1775, (2005).

[30] T. S. Böscke, S. Govindarajan, C. Fachmann, J. Heitmann, A. Avellan, U. Schroeder, S. Kudelka, P. D. Kirsch, C. Krug, P. Y. Hung, S. C. Song, B. S. Ju, J. Price, G. Pant, B. E. Gnade, W. Krautschneider, B-H. Lee and R. Jammy, IEDM, 1-4 (2006)

[31] K. Kim, G. Jeong, IEDM, 27-30 (2007)

[32] G. Lupina, G. Kozlowski, J. Dabrowski, Ch. Wenger, P. Dudek, P. Zaumseil, G.Lippert, Ch.Walczyk and H-J.Müssig, Appl. Phys. Lett. **92**, 062906, (2008)

[33] J. Robertson, J. Vac. Sci.Tech. B 18, 1785, (2000)

[34] M. T. Bohr, R. S. Chau, T. Ghani, and K. Mistry, IEEE Spectrum (2007), see http://spectrum.ieee.org/oct07/5553

[35] T. S. Böscke, S. Govidarajan, P. D. Kirsch, P. Y. Hung, C. Krug, B. H. Lee, J. Heitmann, U. Schröder, G. Pant, B. E. Gnade, and W. H. Krautschneider, Appl. Phys. Lett. **91**, 072902, (2007)

[36] S. Govindarajan, T. S. Böscke, P. Sivasurbramani, P. D. Kirsch, B. H. Lee, H.-H. Tseng, R. Jammy, U. Schröder, S. Ramanathan, and B. E. Gnade, Appl. Phys. Lett. **91**, 062906, (2007)

[37] R. M. Wallace, G. D. Wilk, Materials Issues for High-k Gate Dielectric Selection and Integration in H. R. Huff, D. C. Gilmer (Eds.), "High dielectric constant materials, VLSI MOSFET Application", Advanced Microelectronics, Springer, pp. 261-266, (2005)

[38] R. Shannon, J.Appl.Phys., **73**, 348, (1993)

[39] S. Roberts, Phys. Rev., **76**, 1215, (1949)

[40] I. McCarthy, M. P. Augustin, S. Shamuilia, S. Stemmer, V. V. Afanas'ev, and S. A. Campbell, Thin Solid Films **515**, 2527, (2006)

[41] C. Marchiori, M. Sousa, A. Guiller, H. Siegwart, R. Germann, J.-P. Locquet, J. Fompeyrine, G. J. Norga, and J. W. Seo, Appl. Phys. Lett. **88**, 072913, (2006)

[42] C. Rossel, B. Mereu, C. Marchiori, D. Caimi, M. Sousa, A. Guiller, H. Siegwart, R. Germann, J.-P. Locquet, J. Fompeyrine, and J. Webb, Appl. Phys. Lett. **89**, 053506, (2006)

[43] S. Van Elshocht, J.Vac.Sci.Technol. B 27(1), (2009)

[44] A. Toriumi and K. Kita, in Dielectric Films for Advanced Microelectronics, edited by M. Baklanov, M. Green, and K. Maex Wiley, Chichester, (2007)

[45] H. Ibach, H. Lüth, "Festkörperphysik, Einführung in die Grundlagen", 2 Auflage, Springer Lehrbuch, (1988)

[46] J. Robertson, J. Appl. Phys., **104**, 124111 (2008)

[47] J. Robertson, Journal of non-crystalline solids, **303**, 108-113, (May 2002)

[48] U. Schröder *et al.*, "Do new materials solve the upcoming challenges of future DRAM memory cells?", DPG Spring meeting (2008)

[49] T. Hori, "Gate dielectrics and MOS ULSIs. Principles, Technologies and Applications", Springer, Berlin Auflage (1997)

[50] R. G. Ehl, *et al.*, Journal of Chemical Education **31**, 226–232, (May 1954)

[51] A. Berthelot *et al.*, "Highly Reliable TiN/ZrO$_2$/TiN 3D Stacked Capacitors for 45 nm Embedded DRAM Technologies", Freescale Semiconductor Inc., (2006)

[52] W.J. Zhu *et al.*, IEEE ELECTRON DEVICE LETTERS, **23**, 2, (February 2002)

[53] G. D. Wilk et al., J. Appl. Phys., **89**, 5243 (2001)

[54] P. M. Martin (editor), Handbook of deposition technologies for films and coatings: science, applications and technology, Oxford (2010)

[55] R. Droopad, K. Eisenbeiser, A. A. Demkow, High-k Crystalline Gate Dielectrics: An IC Manufacturer's Perspective in H. R. Huff, D. C. Gilmer (Eds.), "High dielectric constant materials, VLSI MOSFET Application", Advanced Microelectronics, p. 647, Springer, , (2005)

[56] S. Hüfner, Photoelectron Spectroscopy, Springer–Verlag, 2003

[57] J. F. Watts and J. Wolstenholme, *An introduction to Surface Analysis by XPS and AES*, J. Wiley & Sons, West Sussex, 2003

[58] J. F. Moulder, W. F. Stickle, P. E. Sobol, K. D. Bomben, Handbook of X–ray Photoelectron Spectroscopy, Physical Electronics Inc. (1995)

[59] G. Ertl, J. Küppers, "Low energy electrons and surface chemistry", VCH Verlagsgesellschaft mbH (1995)

[60] D. Briggs, M. P. Seah, Practical Surface Analysis by Auger and X–Ray Photoelectron Spectroscopy, Wiley (1990)

[61] G. Lukovsky, J. L. Whitten, Electronic Structure of Alternative High–K Dielectrics, in H. R. Huff, D. C. Gilmer (Eds.), "High Dielectric Constant Materials, VLSI MOSFET Applications", pp. 311-353, Springer (2005)

[62] D. Schmeisser., P. Hoffmann, G. Beuckert, in *Materials for Information Technology, Devices, Interconnects and Packaging, Series: Engineering Materials and Processes*, E. Zschech, C. Whelan, T. Mikolajick (Eds.), Springer, Berlin, 2005, 449–460

[63] D.R. Batchelor, R. Follath, D. Schmeißer, Nuclear Instr. Methods Phys. Res. A 467–468, 470 (2001)

[64] P. Hoffmann, R.P. Mikalo, D. Schmeißer, Solid-State Electron. **44**, 837 (2000)

[65] M. Kitamura, H. Chen, Ferroelectrics, **210**, 13 (1998)

[66] G. Lukovsky, J. G. hong, C. C. Fulton, Y. Zou, R. J. Nemanich, H. Ade, D. G. Scholm, J. L. Freeouf, Phys. Stat. Sol. B, **241**, 2221-2235 (2004)

[67] J. H. Richter, P. G. Karlsson, A. Sandell, J. Appl. Phys. **103**, 094109 (2008)

[68] D. Schmeisser, M. Tallarida, K. Henkel, K. Müller, D. Mandal, D. Chumakov, E. Zschech, Mat. Sc. Poland, **27**, 1 (2009)

[69] O. Seifarth, PhD dissertation BTU Cottbus (2006)

[70] P. Zaumseil, T. Schroeder, J. Phys. D: Appl. Phys. **38**, A179 (2005) and N. Terasawa, K. Akimoto, Y. Mizuno, A. Ichimiya, K. Sumitani, T. Takahashi, X. W. Zhang, H. Sugiyama, H. Kawata, T. Nabatame, A. Toriumi, Appl. Surf. Sci., 244, 16 (2005)

[71] Jens Als-Nielsen, Des McMorrow, *Elements of Modern X-Ray Physics*, Wiley, New York, (2008)

[72] P. Zaumseil, RCRefSim (Rocking Curve and Reflectivity Simulation), IHP Frankfurt Oder (2005)

[73] B. D. Cullity, S. R. Stock, Elements of X–Ray diffraction, Prentice Hall, 3rd edition (2001)

[74] M. Birkholz *Thin Film Analysis by X-Ray Scattering,* Wiley-VCH Verlag (2006)

[75] G. Binnig, H. Rohrer, Ch. Gerber, E. Weibel, Appl. Phys. Lett. **40**, 178 (1982)

[76] G. Meyer, N. M. Amer, Appl. Phys. Lett., **53**, 1045 (1988)

[77] G. Binnig, C. F. Quate, Ch. Gerber, Phys. Rev. Lett. **56(9)**, 930 (1986)

[78] S. N. Magonov, M.-H. Whangbo Surface Analysis with STM and AFM. Experimental and Theoretical Aspects of Image Analysis.VCH (1996)

[79] Keithley Application Note Series, Number 2896 (2009)

[80] G. A. Brown, Electrical Measurements Issues for Alternative Gate Stack Systems, in H. R. Huff, D. C. Gilmer (Eds.), High Dielectric Constant Materials, pp. 521-562, Springer (2004)

[81] J. R. Hauser, K. Ahmed, Characterization of Ultra–Thin Oxides Using Electrical C–V and J–V Measurements, Characterization and Metrology for ULSI Technology (1998)

[82] Nanoscale Phenomena in Ferroelectric Thin Films, ed. S. Hong, Kluwer Academic Publishers (2004)

[83] J.A. Christman, S.H. Kim, H. Maiwa, J.P. Maria, B.J. Rodriguez, A.I. Kingon, R. J. Nemanich, J. Appl. Phys. **87**, 8031 (2000)

[84] Image Processing Software for Microscopy, see: http://www.imagemet.com/

[85] J. H. Scofield, J. Electron. Spectrosc. Relat. Phenom. **8**, 129–137 (1976)

[86] R. F. Reilman, A. Msezane, S. T. Manson, J. Electron Spectrosc. Relat. Phenom. **8**, 389–394 (1976)

[87] M. P. Seah, W. A. Dench, Surf. Interface Anal. **1**, 2–11 (1979)

[88] J.–N. Kim, K.–S. Shin, B.–O. Park, J.–H. Lee, N.–K. Kim, S.–H. Cho, Smart Mater. Struct. **12**, 565–570 (2003)

[89] H. Bubert, J. C. Rivière, Surface and Thin Film Analysis: Principles, Instrumentation, Applications, Ed. H. Bubert, H. Jenett, Wiley–VCH Verlag (2002)

[90] C. D. Wagner, L. E. Davis, M. V. Zeller, J. A. Taylor, R. H. Raymond, L. H. Gale, Surf. Interf. Anal. **3**, 211 (1981)

[91] J. A. T. Verhoeven, H. van Doveren, Appl. Surf. Sci. **5**, 361-373 (1980)

[92] T. L. Barr, J. Va. Sci. Technol. **A 9**, 1793 (1991)

[93] M. H. Hecht, I. Lindau, D. S. Chen, J. Appl. Phys. **53**, 9021 (1982)

[94] L. T. Hudson, R. L. Kurtz, S. W. Robey, D. Temple, R. L. Stockbauer, Phys. Rev. B **47**, 16 (1993)

[95] A. R. West, C. West *Basic Solid State Chemistry* Wiley (1999)

[96] T. H. Büyüklimanli, J. H. Simmons, Phys. Rev. B **44**, 727 (1991)

[97] M. Renninger Zeitschrift für Physik A Hadrons and Nuclei **106**, 141–176 (1937)

[98] T. Maekawa, K. Kurosaki, and S. Yamanaka, J. Alloys Compd. **407**, 44 (2006)

[99] J. Dabrowski, P. Dudek, G. Kozlowski, G. Lupina, G. Lippert, R. Schmidt, Ch. Walczyk, Ch. Wenger ECS Trans., **25(6)** 219-239 (2009)

[100] S. M. Sze, Physics of Semiconductor Devices, 2nd Edition, 402-407, Wiley, (1981)

[101] J. Frenkel Phys. Rev. **54**, 647 (1938)

[102] J. R. Yeargan, H. L. Taylor, J. Appl. Phys. **39**, 5600–5604 (1968)

[103] Z. Xu, M. Houssa, S. D. Gendt, M. Heyns, Appl. Phys. Lett. **80**, 1975 (2002)

[104] H.-J. Noh, B. J. Kim, S.-J. Oh, J.H. Park, H.-J. Lin, C.T. Chen, Y.S. Lee, K. Yamaura, E. Takayama-Muromachi, J.Phys.: Condens. Matter., **20**, 485208 (2008)

[105] M. Merz, N. Nücker, E. Pellegrin, S. Schuppler, M. Kielwein, M. Knupfer, M. S. Golden, J. Fink, C. T. Chen, V. Chakarian, Y. U. Idzerda, A. Erb, J. Low Temp. Phys. **105**, 3/4 (1996)

[106] J. H. Richter, P. G. Karlsson, B. Sanyal, J. Blomquist, P. Uvdal, A. Sandell, J. Appl. Phys. **101**, 104120 (2007)

[107] W. Feller, An Introduction to Probability Theory and Its Applications, vol. 1, 3rd Edition, Wiley (1970)

[108] K. Szot, W. Speier, Phys. Rev. B, **60**, 5909–5926 (1999)

[109] S. A. Campbell, D. C. Gilmer, X. Wang, M. T. Hsich, H. S. Kim, W. L. Gladfelter, J. H. Yan, IEEE Trans. Electron Devices **44**, 104 (1997)

[110] R. B. van Dover, Appl. Phys. Lett. **74**, 3041 (1999)

[111] T. Tohma et al., Jpn. J. Appl. Phys., **41**, 6643 (2002)

[112] G. D. Wilk, R. M. Wallace, J. M. Anthony J. Appl. Phys. Rev. **89**, 5243–5275 (2001)

[113] W. H. Payne, V. J. Tennery, J. Amer. Cer. Soc., **48**, 413-417 (1965)

[114] V. Young, P. C. McCaslin, Anal. Chem. 57, 880 (1985)

[115] C. Chang, H–F. Liu, C–H. Hsieh, J–M. Chen, Phys. Rev. B **60**, 7703 (1999)

[116] M. C. Chang, S. Sugihara, J. Mater. Sci. Lett. **20**, 237–239 (2001)

[117] U. Diebold, Surface Science Reports **48**, 53–229 (2003)

[118] C. Wang, M. H. Kryder, J. Phys. D: Appl. Phys. **41**, 245301 (2008)

[119] J. Xu, J. Zhai, X. Yao, Appl. Phys. Lett. **89**, 252902 (2006)

[120] A. L. Patterson, Phys. Rev. **56**, 978 (1939)

[121] J. T. Last, Phys. Rev. **105**, 1740 (1957)

[122] A. R. Denton, N. W. Ashcroft, Phys. Rev. A **43**, 3161 (1991)

[123] M. Czernohorsky, E. Bugiel, H. J. Osten, A. Fisel, O. Kirfel, Appl. Phys. Lett. **88**, 152905 (2006)

[124] T. Schroeder, P. Zaumseil, G. Weidner, Ch. Wenger, J. Dabrowski, H–J. Müssig, P. Storck, J. Appl. Phys. **99**, 014101 (2006)

[125] K. Eisenbeiser, Device *principles of high–k dielectrics*, in Materials Fundamentals of Gate Dielectrics, Springer (2006)

[126] H. Kim, P. C. McIntyre, K. C. Saraswat Appl. Phys. Lett. 82, 106 (2003)

[127] H. G. Tompkins and E. A. Irene, *Handbook of Ellipsometry*, pp.146–150, Springer (2005)

[128] E. J. Johnson, *Absorption near the fundamental edge* in: *Semiconductors and Semimetals*, vol. 3, Academic Press, New York (1967)

[129] R.A. Smith, *Wave Mechanics of Crystalline Solids*, Chapman and Hall LTD, London (1961)

[130] B. Fromme, "d–d Excitations in Transition–Metal Oxides, A Spin–Polarized Electron Energy–Loss Spectroscopy (SPEELS) Study", 6–7, Springer Verlag (2001)

[131] G. Lupina, O. Seifarth, G. Kozlowski, P. Dudek, J. Dabrowski, G. Lippert, H–J. Müssig, Microelectronic Engineering 86, 1842 (2009)

[132] H. Olin, Meas. Sci. Technol. 5, 976–984 (1994), K. K. Leang, S. Devasia, IEEE Trans. Contr. Syst. Tech. 15(5),927–935 (2007)

[133] G. Lupina, J. Dąbrowski, P. Dudek, G. Kozłowski, P. Zaumseil, G. Lippert, O. Fursenko, J. Bauer, C. Baristiran, I. Costina, H–J. Müssig, L. Oberbeck, U. Schröder, Appl. Phys. Lett. 94, 152903 (2009)